Cogeneration—Combined Heat and Power (CHP)

THERMODYNAMICS AND FLUID MECHANICS SERIES

General Editor: W. A. WOODS

Other Pergamon Titles of Interest

BENSON & WHITEHOUSE
Internal Combustion Engines (2 volumes)

DIXON
Fluid Mechanics, Thermodynamics of Turbomachinery

DIXON
Worked Examples in Turbomachinery (Fluid Mechanics and Thermodynamics)

DUNN & REAY
Heat Pipes, 2nd Edition

GOSTELOW
Cascade Aerodynamics

HAYWOOD
Analysis of Engineering Cycles, 3rd Edition

HAYWOOD
Analysis of Engineering Cycles, Worked Problems

LAI *et al*
Introduction to Continuum Mechanics, Revised Edition

RAO
The Finite Element Method in Engineering

REAY & MACMICHAEL
Heat Pumps

Pergamon Related Journals

(*Free specimen copy gladly sent on request*)

International Communications in Heat and Mass Transfer
International Journal of Heat and Mass Transfer
International Journal of Mechanical Sciences
International Journal of Multiphase Flow
Journal of Heat Recovery Systems
Mechanism and Machine Theory

Cogeneration—Combined Heat and Power (CHP)

Thermodynamics and Economics

by

J. H. HORLOCK, F.Eng., F.R.S.

Vice-Chancellor, Open University

PERGAMON PRESS
OXFORD · NEW YORK · BEIJING · FRANKFURT
SÃO PAULO · SYDNEY · TOKYO · TORONTO

U.K.	Pergamon Press, Headington Hill Hall, Oxford OX3 0BW, England
U.S.A.	Pergamon Press, Maxwell House, Fairview Park, Elmsford, New York 10523, U.S.A.
PEOPLE'S REPUBLIC OF CHINA	Pergamon Press, Room 4037, Qianmen Hotel, Beijing, People's Republic of China
FEDERAL REPUBLIC OF GERMANY	Pergamon Press, Hammerweg 6, D-6242 Kronberg, Federal Republic of Germany
BRAZIL	Pergamon Editora, Rua Eça de Queiros, 346, CEP 04011, Paraiso, São Paulo, Brazil
AUSTRALIA	Pergamon Press Australia, P.O. Box 544, Potts Point, N.S.W. 2011, Australia
JAPAN	Pergamon Press, 8th Floor, Matsuoka Central Building, 1-7-1 Nishinshinjuku, Shinjuku-ku, Tokyo 160, Japan
CANADA	Pergamon Press Canada, Suite No. 271, 253 College Street, Toronto, Ontario, Canada M5T 1R5

First edition 1987

Library of Congress Cataloging in Publication Data

Horlock, J. H.
Cogeneration—Combined heat & power (CHP).
(Thermodynamics and fluid mechanics series)
Includes bibliographies.
1. Cogeneration of electric power and heat.
I. Title. II. Title: Combined heat and power.
III. Series.
TK1041.H67 1987 621.31′21 86-25218

British Library Cataloguing in Publication Data

Horlock, J. H.
Cogeneration—Combined heat & power (CHP)—
(Thermodynamics and fluid mechanics series)
1. Heating from central stations 2. Electric
power-plants 3. Waste heat
I. Title II. Series
697′.54 TH7641

ISBN 0 08 034797 5 Hardcover
ISBN 0 08 034796 7 Flexicover

Printed in Great Britain by A. Wheaton & Co. Ltd., Exeter

Editorial Introduction

The books in the Pergamon Thermodynamics and Fluid Mechanics Series were originally planned as a set to cover the subjects taught in three-year undergraduate courses for Mechanical Engineers. Subsequently the aims of the series were broadened and several volumes were introduced which catered not only for undergraduates but, also, for postgraduate students and engineers in practice. These included new editions of books published earlier in the series.

This book, written by Dr. Horlock, continues the policy of broadening the scope of the series and its publication is particularly appropriate and timely. Increasing importance is being given in engineering education to the study of engineering applications and economics as a result of the policies of the U.K. Engineering Council. This volume will be useful for design and development engineers in industry. It will help students on Master's degree courses and final year honours undergraduates, particularly those studying power plants and energy.

Dr. Horlock's work includes a rigorous, but not over-pedantic, treatment of thermodynamic processes and interactions. The approach adopted is based upon the laws of thermodynamics as basic postulates. The concept of internal energy results from the first law and that of entropy from the second law. Heat and work are interactions which, in general, produce changes in the internal energies of both the system and the surroundings. The expressions "heat added to the system" and "work done by the system" are used as acceptable descriptions of interactions which result in a change of internal energy of the system. Interactions are also considered as time–rate events and expressions such as 'the rate of heat transfer', 'heat rate' and 'rate of working' or 'power' are accepted. The title "Combined heat and power" mixes an interaction with a rate of interaction but, notwithstanding this, it has become established convention, as the author points out.

1987   W. A. WOODS

Contents

Chapter 4   Economic Assessment of CHP Schemes   **111**

Notation

Note: Lower case symbols for properties represent specific quantities (i.e. per unit mass).

Symbol	Meaning	Typical Units
A	annual cash flow	£, $ p.a.
b; B	steady-flow availability	kJ/kg; kJ
c, C	unit capital cost; capital cost	£, $/kW; £, $
ΔC	extra capital cost (e.g. for cogeneration plant)	£, $
(CP)	coefficient of performance (of heat pump)	(—)
(CV)	calorific value	kJ/kg
D	housing density	number/acre
DCFRR	discounted cash flow rate of return	(year)$^{-1}$
e; E	exergy	kJ/kg; kJ
EUF	energy utilisation factor	(—)
f; F	fuel energy supplied	kJ/kg; kJ
$_{N}f_{AP}$	present worth factor	(—)
$_{N}f_{AF}$	future worth factor	(—)
f_{N}	capitalised cost factor	(—)
FESR	fuel energy savings ratio	(—)
g, G	Gibbs function	kJ/kg; kJ
h, H	enthalpy	kJ/kg; kJ
H	plant utilisation	h/year
i	interest or discount rate	(—)
IHR	incremental heat rate	(—)
K	heat load (density)	MW/m^2
l, L	delay periods, periods of investment	years
h_{fg}	latent heat	kJ/kg
m	mass; mass of bled steam (per unit boiler flow); operational factor	kg; (—); (—)

M	annual fuel costs; mass of circulating steam (per unit boiler flow)	£, \$ p.a.; (—)
N	plant life	years
OM	annual operational and maintenance costs	£, \$ p.a.
P	cost per year	£, \$ p.a.
PBP	pay-back period	(years)
q; Q	heat supplied or rejected	kJ/kg; kJ
ROI (DCFRR)	investor's rate of return	$(\text{year})^{-1}$
S	fuel cost per unit mass	£, \$/kg
$\mathscr{S}$	fuel cost per unit energy	£, \$/kWh
s, S	entropy	kJ/kgK; kJ/K
(SALV)	salvage value	£, \$
t	time; tax rate or credit	s; (—)
w, W	work output	kJ/kg, kJ
x	dryness fraction of steam	(—)
Y	"unitised" price (price per unit of energy)	£, \$/kWh
z	fraction (of capital)	(—)
Z	Z-factor (lost work/useful heat supplied)	(—)
α, β	areas on T,s diagram—enthalpy differences—as defined in text	kJ/kg
$\beta = {}_N f_{AF}$	capitalised cost factor	(—)
γ	non-dimensional fuel saving	(—)
η	efficiency	(—)
ϕ	loss factor in transmission	(—)
ψ	performance criterion for waste heat boiler	(—)
λ	heat/work ratio	(—)
ϵ	heat exchanger effectiveness	(—)

Subscripts

a	annual; artificial
a, b, c, d	stages in conversion
A	relating to heat supply
B	boiler; relating to heat supply
BI	bought-in (from utility)
BB	buy-back or sale (to utility)
c	compressor
cc	combustion

C	conventional plant
CAR	Carnot
CD	credit for depreciation
CI	credit for investment
CG	cogeneration (CHP) plant
CV	control volume
d	debt
D	demand
DCF	discounted cash flow
E	electrical
eq	equivalent; equity
f	fuel; fluid (water)
G	gas
H	higher (upper, topping)
H	heat (price)
HP	heat pump
L	lower (bottoming)
NU	rejection of non-useful heat
o	overall
ON	on peak
OFF	off peak
P	product, pump
R	reactant, rational
REF	reference
S	space (heating); stack
T	turbine
TH	thermal
U	rejection of useful heat
1, 2, 3 . . .	miscellaneous, but usually referring to states
0	conceptual environment (ambient) state

Superscripts

· (e.g. $\dot{m}$, $\dot{Q}$, $\dot{W}$)	*rate* of (mass flow, heat supply, work output)
′ (e.g. η', $(i)'$)	changed, replacement, or new value (of efficiency, of interest rate, etc.)
⁻ (e.g. $\bar{A}$)	levelised (cash flow); mean or average

Author's Preface

This is a book for engineers. First it involves the use of straightforward thermodynamics to predict the efficiency and energy utilization of combined heat and power (CHP) plant. But the science of thermodynamics is not sufficient to assess the complete performance of such CHP stations; it is the assessment of the possible economic benefits that critically controls whether such projects should go ahead. So I have entered the field of economics to write the later part of this book.

In 1974 I was appointed by the Secretary of State for Energy to the Combined Heat and Power Group, which examined the future role for CHP in the United Kingdom. Membership of that group, under the chairmanship of Dr. Walter (now Lord) Marshall, introduced me to this complex area. I make frequent reference in the text to various reports we produced, to the literature we studied, to work we initiated and to work we undertook ourselves.

I am grateful to many people who have helped me, particularly in reading and commenting on first, second and subsequent drafts. In the academic world, I have been much influenced by Mr. R. W. Haywood, my teacher when I was a student, my colleague when I worked in the Cambridge University Engineering Department and my friend over many years. In the early chapters I have many times referred back to one of the earlier books in this series, *Analysis of Engineering Cycles*, by Haywood. His precision in both engineering science and the English language has established high standards for any engineer to emulate. My colleague at the Open University, Professor L. Harris, has commented helpfully upon my attempts at elementary economics.

In the industrial world, I have benefited from a long association with engineers of the General Electric Company, including two of my former students, Dr. H. J. Perkins and Dr. D. Pollard. Mr. P. D. Lilley and Mr. J. Parsons have also been most helpful, particularly in commenting on Chapters 2 and 3. From the Atomic Energy Research Establishment at Harwell, Dr. G. Hewitt and Dr. G. Owen have given assistance with comments on Chapter 4, and for assistance with information on the practical CHP schemes of Chapter 5, I am grateful to the following:

Messrs. J. Bitterli, H. J. Leimer and W. Ruzek, of Sulzers;
Mr. J. D. Lycett, of the Midlands Electricity Board; and
Mr. D. V. Gray, of the Open University.

But none of those I have listed bears any responsibility for errors or interpretations in this book.

The life of a Vice-Chancellor is a strange, "grass-hopper" type of existence. To maintain an academic and scientific status, as I have attempted to do, he has to accept that time for a steady sustained effort on a particular project is difficult to find. The best to be hoped for is an hour or two here and there, a weekend or a short period of study leave away from the heavy administrative load and the political battles associated with U.K. higher education in the 1980s. Inevitably this means that one's "personal" time has to be used extensively, which imposes a strain on family life. I am conscious of the patience and understanding that my wife has shown in helping me to remain active in the scientific world.

Further, I am much indebted to Mrs Sheila Watts, who has deciphered many draft manuscripts, transforming them into finished and highly respectable typescripts for the publisher, Pergamon Press. I should also like to thank Mr. C. J. Hunter-Brown for considerable help in tracing references.

Finally, it is a pleasure to contribute this book to the Pergamon Thermodynamic and Fluid Mechanics Book Series. I started the series as editor in the sixties, later handing over to my former colleague at Liverpool University, Professor W. A. Woods. I am grateful to him for his help and encouragement.

Introduction

The whole pattern of supply and consumption of energy changed following the very rapid increase in oil prices in 1973 and 1974. Figure I.1 (taken from refs. 1 and 2) shows how world, US and UK primary fuel consumption varied over this period. For the United Kingdom it appeared in the mid-1970s that an "energy-gap"—the difference between consumption and the total indigenous production of fuel—might develop. In fact economic downturn and energy savings have led to a reduction in consumption and such a gap now appears unlikely to develop in the United Kingdom in the twentieth century. Traditional energy sources will continue to meet energy demand with little contribution from "renewable"

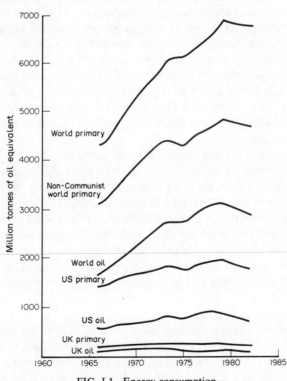

FIG. I.1. Energy consumption.

TABLE I.1. *UK Primary Energy Consumption*

| | Million tonnes of oil equivalent | | | | |
	1963	1973	1983	1990	2000
				←Projected→	
Oil	61	114	72	69	64
Natural gas	—	26	44	51	55
Coal	116	78	66	61	69
Nuclear, etc.	3	7	12	18	21
Total	180	225	194	199	209

1. Includes non-energy use of oil and natural gas.
2. "Coal" includes other coal derived fuels.
3. "Nuclear etc." includes hydropower and any other new sources (e.g. wind, solar).

sources of supply, although significant changes in the form of supply have taken place, as Table I.1 (taken from ref. 3) indicates.

For industrial countries such as the United Kingdom, several policy options are available to bridge an energy gap should it begin to develop again. These options include increased production of coal, the exploration for more oil and gas, the use of renewable resources of energy (wind, waves, tides, geothermal sources), the further development of nuclear power, and the general conservation of energy. Within the latter there are many ways of achieving the desired result. There is improved thermal insulation, coupled with use of heat pumps; improvement in the efficiency of heat sources (e.g. domestic boilers); the use of waste products as fuel; and the option with which this book is concerned, that of combining heat and power production.

Power plants involve the production of electricity from a source of energy. Conventional "thermal" power plants use the chemical energy in a hydrocarbon fuel or the energy of fission in a nuclear fuel as the source. Renewable sources used in less conventional power plants include potential energy (in hydroelectric schemes), kinetic energy (in wind power generators), or a supply of working fluid with high stagnation enthalpy (in geothermal power plants). Clearly the replacement of thermal power stations by plants using renewable sources will save "expendable" energy. But so will the use of combined heat and power plant, in which the energy supplied to a thermal power station is more effectively used to produce both electricity and useful heat.

In conventional thermal power stations only about one-third of the energy in the coal or oil appears as electrical power—two-thirds of the energy is thrown away in the form of lukewarm water, in cooling towers, to rivers or to the sea. Alteration of the design and operation of an electrical power station to cogeneration—the production of both useful heat and work—improves the energy utilisation. The heat must be provided at a sufficiently high temperature for space and hot water requirements of domestic, commercial and public buildings, or alternatively steam may be provided to meet industry's needs for processes.

These two general areas of cogeneration (combined heat and power for district heating (CHP/DH) and combined heat and power for industry (CHP/IND)) have been the subject of much detailed work during the past few years. The essence of the first (CHP/DH) involves the construction of district heating networks which supply heat for space and hot water at water temperatures in the range 80–150°C. In CHP/IND (e.g. for factories requiring process heat) it is more usual for the heat demand to be met by steam exhausting or bled from a steam power turbine (or sometimes raised in a boiler fed by gases from an industrial gas turbine).

We do not include in this definition of combined heat and power the use of the warm water which is rejected from existing power stations, for agricultural and fish farming. Nor do we include the use of waste incineration as fuel for an electrical power station. Essentially, we consider here the modification of existing power plants or the design of new ones to provide the rejected heat in a useful form (i.e. at higher temperatures than in conventional plants).

Although there are many small district heating schemes in the United Kingdom and the United States, and substantial use of process heat supplied from special power stations, the overall use of CHP/DH is relatively limited. But such cogeneration for district heating is used more widely in several European countries (see ref. 4). For example, in Denmark about 25% of the total heat load is supplied through district heating, of which one-third comes from CHP plant (8%); about 20% of Swedish space and water heating is supplied through district heating schemes, of which about three-fifths (i.e. 12%) comes from CHP plant; about 0.7% of West German space and water requirements is met by CHP plant. In Finland, there is only one major town *without* district heating and at the end of the seventies 17% of the heat to buildings was met from district heating (with a planned increase to 40% in the eighties), about half produced by CHP plant.

The energy savings that result from combined heat and power are substantial. The thermodynamics of CHP plant are relatively straightforward and are described in detail in the early chapters of this book. This basic study of thermodynamics leads to accurate assessment of energy savings, provided component efficiencies within the power plant are known accurately. However, the matching of the plant to required heat and electrical power loads is often complex; further, it is often the economics of combined heat and power operation which has held up the development of this policy option for energy saving. This aspect of combined heat and power merits the detailed study which is given in this volume.

TABLE I.2. *Renewable Resources*

---

12 mtoe per year (6% of UK primary energy consumption in 1990) could be saved:
  (i) If 5 GW of electrical power were produced *continuously*
      from the Severn Barrage (tidal power)
      or from 1250 4 MW windmills (each 60 m dia)
      or from wave power "rafts" (300 km long)
      or from solar panels (20 km × 20 km)
 (ii) If 38 GW (30% of the UK heat load) were met by CHP plant, instead of
      using the existing mix of fuel supplies.

The potential for energy savings through combined heat and power in the United Kingdom have been given in detail by the Combined Heat and Power Group appointed by the Secretary of State for Energy in 1976 and which reported in 1979 (ref. 4). The group concluded that if CHP could capture 30% (about 38 GW) of the total UK heat load (128 GW), the approximate savings compared with the existing fuel mix in the United Kingdom approached 12 million tons of oil equivalent. Table I.2 (taken from ref. 1) gives other alternatives for saving or providing about 6% of the UK energy bill. Combined heat and power therefore remains a sound policy option for energy conservation in the United Kingdom and in other countries and it merits detailed study by students, postgraduates, and practising engineers. It is the basis for that study which this book provides.

References

1. Horlock, J. H. *Energy Resources*. Royal Society/British Association. The Sir David Martin Lecture, 1980. Sri Lankan Association for the Advancement of Science Lecture, 1983.
2. B.P. *Statistical Review of World Energy*, 1982.
3. Esso Magazine Supplement. *United Kingdom Energy Outlook*, 1984/85.
4. *Combined Heat and Electrical Generation in the United Kingdom* (The Marshall Report). Department of Energy, Energy Paper No. 35, HMSO, 1979.

CHAPTER 1

Thermodynamics of Conventional Power Producing Plants—A Brief Review

1.1 Introduction

The thermodynamics of thermal power plants has long been a classical area of study for engineers. The objectives of such work have traditionally been the determination and maximisation of thermal efficiency, i.e. the most efficient production of (usually electrical) power from a supply of fuel with chemical energy. Figure 1.1(a) shows a block diagram of a conventional power plant (C) receiving fuel energy (F_C), producing work (W_C) and rejecting "non-useful" heat $(Q_{NU})_C$ to a sink at low temperature. The designer attempts to minimise the fuel input for a given work output because this will clearly give economic benefit in the operation of the plant, minimising fuel costs against the sales of electricity to meet the power demand. However, the capital cost of achieving higher thermal efficiency has to be assessed and balanced against resulting savings in fuel costs.

The objectives of the designer of combined heat and power plant are wider—both heat and work production. Figure 1.1(b) shows a block diagram of a CHP or cogeneration (CG) plant receiving full energy (F_{CG}) and producing work (W_{CG}). But now useful heat $(Q_U)_{CG}$ is now produced, as well as non-useful heat $((Q_{NU})_{CG})$. Both the work and the useful heat can be sold, so the CHP designer is not solely interested in high thermal efficiency, although the work output commands a higher sale price than the useful heat output. Clearly both thermodynamics and economics will be of importance.

But before studying CHP plant we briefly review the thermodynamics of conventional power plant, as we shall build on that science and use similar concepts and notation (the book by Haywood[1]* provides that base). It is important first to distinguish between a power plant and a cyclic heat engine (steady flow type). In the latter, fluid passes continuously round a closed circuit, through a thermodynamic cycle in which heat (Q_B) is received from a source at high temperature, heat (Q_A) is rejected to a sink at low temperature, and a work output (W) is delivered, usually to drive an electric

---

*Superscript numbers are to references at the end of the chapter.

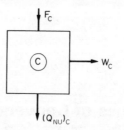

(a) Conventional power plant

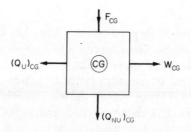

(b) CHP or cogeneration plant

FIG. 1.1. Conventional and CHP plants.

generator. Thus in Fig. 1.2, which shows a steam power plant, the part within the dotted control surface (Y) is a cyclic heat engine (or cyclic heat power plant). Heat is transferred to the boiler tubes, which are within the cyclic heat engine, but a second "open" control surface (Z) surrounds the boiler furnace. This receives air and fuel, and discharges flue gases; they are obviously not recirculated, so the boiler furnace is not a cyclic plant. The two control volumes—one closed, the other open—together form a steam power plant.

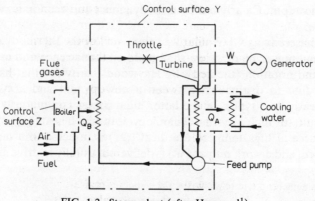

FIG. 1.2. Steam plant (after Haywood[1]).

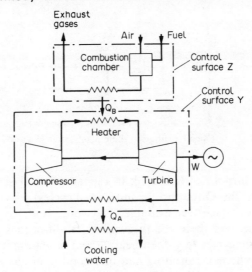

FIG. 1.3. Closed circuit gas turbine
plant (after Haywood[1]).

A gas turbine power plant may similarly operate on a closed circuit (see Fig. 1.3). Again the dotted control surface (Y) surrounds a cyclic heat engine (or cyclic heat power plant) through which air or gas circulates, and the combustion chamber is located within the second, open control surface (Z). The two control volumes form a gas turbine power plant.

However, more usually a gas turbine plant operates on "open circuit", with internal combustion (Fig. 1.4). Fuel and air enter the single control volume and exhaust products leave—they are not recycled. This open-circuit plant cannot be said to operate on a thermodynamic cycle; however, its performance is often assessed by treating it as equivalent to a closed-circuit cyclic power plant, but care must be used in such an approach. Internal combustion engines similarly operate on "open circuit".

The classical heat engine cycles for power production in steam and gas turbine plant are those associated with the names of Rankine and Joule-Brayton respectively, and the temperature–entropy diagrams for these cycles are shown in Figs. 1.5 and 1.6. The Rankine cycle is the basis of cyclic steam power plant, with steady flow through a boiler, a turbine driving a generator delivering electrical power, a condenser and a feed-pump. The Joule-Brayton constant pressure cycle is the basis of the cyclic gas turbine power plant, with steady flow of air (or gas) through a compressor, heater, turbine, cooler within a closed circuit. The turbine drives both the compressor and a generator delivering electrical power.

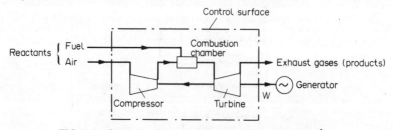

FIG. 1.4. Open circuit gas turbine plant (after Haywood[1]).

It is not the purpose of this book to repeat the analyses of these and other cycles (e.g. the Otto, Stirling and Diesel cycles) which are available in many textbooks (see, for example, Haywood[1] and Rogers and Mayhew[2]). Also described there are the various modifications which may be made to the basic cycles (e.g. feed heating and reheating in steam cycles; internal heat exchange, reheating and intercooling in gas turbine cycles) to increase thermal efficiency. The purpose here is to study the thermodynamics of plant designed with different objectives—the production of

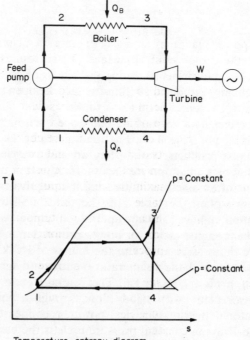

FIG. 1.5. Rankine cyclic heat engine.

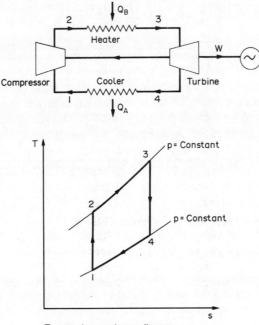

Temperature-entropy diagram

FIG. 1.6. Joule–Brayton cyclic heat engine.

both heat *and* power—referred to as combined heat and power (CHP) plant. But many of the basic thermodynamic ideas developed for conventional work-producing plant are obviously carried across to this different field of study, even if the criteria of performance are changed. For example, the development of the concepts of availability and exergy (see Haywood[3] and Kotas[4]) are important when maximum work output is the objective. This theory, which gives the maximum work output that can be achieved in steady flow between two stable states 1 and 2 in the presence of a conceptual environment (e.g. the atmosphere) at temperature T_0, remains valid, but needs reinterpretation if heat rejection to a prescribed heat "load" is also required in the process.

One particularly important field of study for conventional power plant is that of the "combined plant" (see ref. 5, for example). A broad definition of the conventional "combined cycle" is one in which a "higher" (upper or topping) thermodynamic cycle produces power, but part or all of its heat rejection goes to *supply* heat to a "lower" (or bottoming) cycle. (In practice the "upper" plant is often open circuit, not cyclic.) The objective is a greater work output for a given heat (or fuel energy) supply, i.e. a greater overall efficiency. A CHP plant is not a "combined" plant in the strict sense of such a definition; the objective is not simply to increase work output, but to achieve both heat and work output, often in a pre-scribed ratio. However, again the thermodynamic analyses that have been developed for combined (power) plant have some relevance for combined heat *and* power plant, and we shall review them later in this chapter.

The term "cogeneration" is often used, particularly in the United States, when both power and useful heat are being produced. It is also sometimes used to describe combined *power* plant, but we shall not use it in that way here. Thus in this text "cogeneration plant" may be used only as an alternative to "CHP plant", although some such plants may indeed involve combined cycles.

1.2   Criteria for Performance of Power Plants

1.2.1   Efficiency of a Closed Circuit Plant

For a cyclic power plant in which fluid circulates continuously within the plant (e.g. the plant enclosed within the control surface Y in Fig. 1.2), the criterion of performance is simply the thermal or cycle efficiency,

$$\eta_{TH} = \frac{W}{Q_B}, \tag{1.1}$$

where W is the net work output and Q_B is the heat supplied. W and Q_B may be measured over a given period of time or per unit mass of the fluid circulating. The efficiency may thus also be expressed in terms of the power output ($\dot{W}$) and the rate of heat transfer ($\dot{Q}_B$),

$$\eta_{TH} = \frac{\dot{W}}{\dot{Q}_B}, \tag{1.2}$$

and this formulation is sometimes more convenient for a steady flow cycle. In most of the thermodynamic analysis (in this chapter and in Chapters 2 and 3), we shall work in terms of W, Q, etc. (kJ/kg), rather than in terms of the rates $\dot{W}$, $\dot{Q}$, etc. (kW), with the objective of achieving simplicity. How-ever, it becomes necessary to use rates ($\dot{W}$, $\dot{Q}$) when we move to economic analyses (Chapter 4) and the description of practical CHP plants (Chapter 5).*

---

*A minor problem of nomenclature arises from these definitions. Equation (1.1) defines a performance criterion based on heat and work quantities, but equation (1.2), although leading to an identical criterion for plant in which flows are steady, involves the *rate* of heat supplied and the *rate* of work produced (power). Strictly we should refer to a combined heat and *work* plant, or a combined heat (rate) and power plant. But it has become a convention to link *heat* and *power* together in referring to CHP plant and we cannot but follow that convention here. Reference to cogeneration plant avoids this difficulty.

The heat supply to the cyclic power plant of Fig. 1.2 comes from the control surface Z. Within this second control surface, a steady flow heating device is supplied with reactants (fuel and air) and discharges products of combustion, and we may define a second efficiency for the "heating device" (or boiler),

$$\eta_B = \frac{Q_B}{m_f(CV)_0} = \frac{Q_B}{F}. \tag{1.3}$$

Q_B is the heat transfer from Z to the closed cycle within control surface Y, which occurs during the time interval when m_f is the mass of fuel supplied; and $(CV)_0$ is its calorific value per unit mass. $F = m_f(CV)_0$ is equal to the heat transfer from Z if the products were to leave this control surface at the entry temperature of the reactants, taken as the temperature of the environment, T_0.

The overall efficiency of the entire plant, including the cyclic power plant (within Y) and the heating device (within Z) is given by

$$\eta_0 = \frac{W}{F} = \left(\frac{W}{Q_B}\right)\left(\frac{Q_B}{F}\right) = \eta_{TH}\eta_B. \tag{1.4}$$

1.2.2 Efficiency of an Open Circuit Plant

For an open circuit (non-cyclic) plant (Fig. 1.4) a different criterion of performance is used—the *rational efficiency* (η_R). This is defined as the ratio of the actual work output to the maximum (reversible) work output that could be achieved between reactants, each at the pressure (p_0) and temperature (T_0) of the environment, and products each at the same p_0, T_0. Thus

$$\eta_R = \frac{W}{W_{REV}} \tag{1.5a}$$

$$= \frac{W}{(-\Delta G_0)}, \tag{1.5b}$$

where $\Delta G_0 = (G_P)_0 - (G_R)_0$ is the change in Gibbs function (from reactants to products). The Gibbs function is $G = (H - TS)$, where H is the enthalpy and S the entropy.

ΔG_0 is not readily determinable, and frequently an *(arbitrary) overall efficiency* is defined as

$$\eta_0 = \frac{W}{(-\Delta H_0)} = \frac{W}{m_f(CV)_0} = \frac{W}{F}, \tag{1.6}$$

where $\Delta H_0 = (H_P)_0 - (H_R)_0$, the change in enthalpy from reactants to products, at the temperature of the environment. This is often also referred to as a thermal efficiency, it being implied that there is a heat supply $m_f(CV)_0$ to an "equivalent" closed circuit plant. For many reactions ΔH_0 is not substantially different from ΔG_0 (see ref. 6), and, following Haywood[1]), we therefore refer to η_0 as the (arbitrary) overall efficiency.

In this book we shall use the criteria of thermal efficiency, heating device (or "boiler") efficiency, and their product (overall efficiency) or arbitrary overall efficiency for conventional power plant. In Chapter 2 we shall develop other criteria for CHP plant. Further, we shall be basing our analyses of performance upon the assumption of steady flow. Some internal combustion (Diesel or gas) engines have been used in combined heat and power plant. For such plant the steady-flow analyses presented remain valid on a "quasi-steady" assumption: that overall efficiencies are used for equivalent steady flow devices producing the same work output over the period of the engine cycle, and absorbing the same heat supply (or fuel energy input) in that time.

1.2.3 Heat Rate

As an alternative to the thermal or cycle efficiency of equation (1.1), the cycle heat rate (the ratio of heat supply rate to power output) is sometimes used:

$$\text{Heat rate} = \frac{\dot{Q}_B}{\dot{W}} = \frac{Q_B}{W}. \tag{1.7}$$

It is clearly the inverse of the thermal or cycle efficiency, when Q_B and W are expressed in the same units.

Sometimes a "heat" rate is used which is based on energy supplied in the fuel. It is then defined as

$$\text{"Heat" rate} = \frac{m_f(CV)_0}{W} = \frac{F}{W}, \tag{1.8}$$

which is clearly the inverse of the overall efficiency of the closed-circuit plant, as defined in equation (1.4), or the inverse of the (arbitrary) overall efficiency of the open-circuit plant, as defined in equation (1.6).

1.2.4 Heat Pumps

So far, we have discussed power plant, plant designed to produce work *output*. A heat pump uses a work *input* to pump heat from a low temperature to a high temperature, and usually qualifies for description as a cyclic heat engine. A complete combined heat and power plant may include a heat pump, driven by a conventional power plant (Fig. 1.7(a)). A large

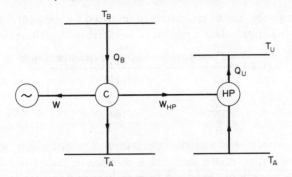

FIG. 1.7(a). Heat pump (HP) driven by
conventional power plant (C).

heat pump is usually a closed circuit plant and operates on a steady flow,
vapour–compression cycle (see Fig. 1.7(b)). A wet liquid–vapour mixture is
fully evaporated at low temperature and pressure, compressed to a super-
heated state, condensed at the higher pressure (rejecting useful heat (Q_U)
to meet the heat load) and throttled to the lower pressure to complete the
cycle. Essentially the cycle is the same as that of a refrigeration cycle, but
here the interest is in pumping heat from the low (usually ambient)

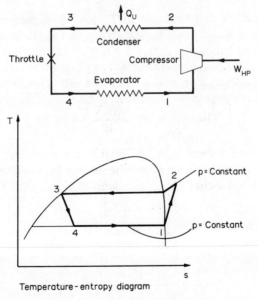

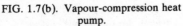

Temperature - entropy diagram

FIG. 1.7(b). Vapour-compression heat
pump.

temperature to the higher temperature required by the heat load, with the minimum work input. The criterion of performance is therefore

$$(CP) = \frac{\text{Heat supplied at upper temperature}}{\text{Work input}}$$

$$= \frac{Q_U}{W_{HP}}, \tag{1.9}$$

where (CP) is referred to as the coefficient of performance, greater than unity. The objective of the designer is to achieve maximum (CP), i.e. to minimise the work input required to meet a specified heat load.

1.3   Ideal (Carnot) Power Plant Performance

The Second Law of Thermodynamics may be used to show that a heat engine achieves maximum efficiency by operating on a reversible cycle called the Carnot cycle, for given (maximum) temperature of supply (T_B), and given (minimum) temperature of heat rejection (T_A). Such a Carnot engine receives *all* its heat (Q_B) at T_B and rejects *all* its heat (Q_A) at T_A (see the temperature–entropy diagram of Fig. 1.8). Its thermal efficiency is

$$\eta_{TH} = \frac{W}{Q_B} = \frac{Q_B - Q_A}{Q_B} = \frac{T_B \Delta s - T_A \Delta s}{T_B \Delta s} = \frac{T_B - T_A}{T_B}. \tag{1.10}$$

Clearly raising of T_B and lowering of T_A lead to higher efficiency. The Carnot engine is a useful hypothetical device in the study of the thermodynamics of power producing cycles, for it provides a measure of the *best* performance that can be achieved under the given boundary conditions on temperature. It is less useful in the study of CHP plant, where both work and heat output are important. However, in Chapter 2 we shall use Carnot

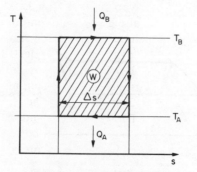

FIG. 1.8. Temperature-entropy diagram for Carnot power plant.

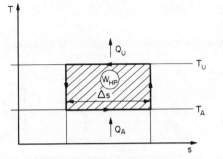

FIG. 1.9. Temperature-entropy
diagram for Carnot heat pump.

CHP engines in analysis leading to the selection of optimum plant for given heat and work loads.

The coefficient of performance of a heat pump operating on a "reversed" Carnot cycle (Fig. 1.9) is

$$(\text{CP}) = \frac{T_U}{(T_U - T_A)},\tag{1.11}$$

where T_U is the temperature of heat delivery (Q_U) and T_A the temperature of the sink from which heat (Q_A) is abstracted. This is the maximum (CP) that could be achieved by a heat pump operating between the temperature levels T_U and T_A.

1.4 The Attainment of High Thermal Efficiency in Conventional Power Plant

1.4.1 Comparison with the Carnot Power Plant

There are two features of the Carnot power plant which give it maximum thermal efficiency:

   (i) all processes involved are reversible;
   (ii) all heat is supplied at the maximum (specified) temperature (T_B) and all heat is rejected at the lowest (specified) temperature (T_A).

In attempting to achieve high thermal efficiency, the designer of a conventional power plant attempts to raise T_B and lower T_A, and to emulate these features (i) and (ii) of the Carnot cycle.

Figure 1.10 shows how this is done for the (Rankine) steam power plant. Firstly, lower rejection temperature (T_A) is achieved by lowering condenser pressure, and it is an important feature of the cycle that all heat is rejected

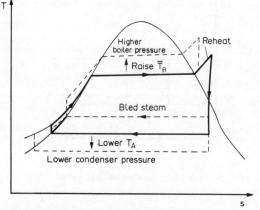

Temperature-entropy diagram

FIG. 1.10. Modifications to basic Rankine
cycle to increase thermal efficiency.

at the lowest temperature. Secondly, efforts are made to raise the mean
temperature of heat supply (T_B) by

(i) raising boiler pressure;
(ii) reheating between turbine cylinders;
(iii) feed heating (bleeding steam from the turbines to heat the condensate before it is fed to the boiler plant).

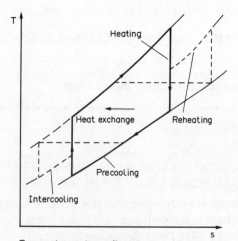

Temperature-entropy diagram

FIG. 1.11. Modifications to basic
Joule–Brayton cycle to increase
thermal efficiency.

Such additional steps must be taken without the introduction of substantial irreversibilities.

Figure 1.11 shows how efficiency can be increased in a closed gas turbine plant by modification of the basic Joule-Brayton cycle, again by raising the mean temperature of heat supply (by reheating) and also by lowering the mean temperature of heat rejection (by intercooling). Further, the introduction of a heat exchanger, to enable the turbine exhaust to heat the compressed gas, achieves a higher mean supply temperature and a lower heat rejection temperature simultaneously. (It is of interest to note that a long-established, successful CHP plant at Coburg in Germany (see Bammert[7]) is based on this advanced cycle. The pre-cooling and inter-cooling phases are partially used to supply district heating, and some 67% of the input energy is utilised).

Whether such modifications to achieve increased thermal efficiency justify the extra capital required is a matter for economic study; thermodynamics is not always the sole criterion. In his search for higher efficiency, the designer of the conventional power plant is maximising the work output (W) for a given heat supply (Q_B). In doing so he is minimising the "non-useful" heat rejection from the plant (Q_A), for, from the First Law of Thermodynamics, $Q_A = (Q_B - W)$. The designer of CHP plant is interested in using some of the heat rejected usefully (Q_U) (so that $Q_A = Q_U + Q_{NU}$, where Q_{NU} is "non-useful" heat rejected) and he may not therefore place emphasis on maximising thermal efficiency. We shall consider alternative criteria of performance for CHP plant in the next chapter. However, because electric power is a valued product (of greater value than useful heat rejected), the CHP plant designer may still wish to produce that electricity at minimum cost, so it is important for him to be aware of the steps taken to maximise conventional plant efficiency. Further, he will be subject to economic limitations, as is the designer of conventional plant, being required to justify capital expenditure to gain thermodynamic performance.

1.4.2 Combined Power Plant

A further development in the search for high thermal efficiency of conventional power plant has been the introduction of combined plant—a subject of intensive study in recent years. Essentially, the use of combined power plant involves seeking higher temperatures of heat supply and lower temperatures of heat rejection.

Consider a cyclic power plant, made up of two cyclic plants (H, L) combined (see Fig. 1.12). Heat rejected (Q_{HL}) from the "higher" (topping) plant (H), of thermal efficiency η_H, is used as a supply to the lower (bottoming) plant L, (of thermal efficiency η_L). The two plants are closed circuit

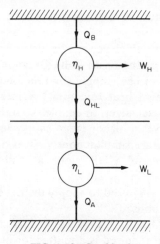

FIG. 1.12. Combined
power plant.

and cyclic, and may use different fluids. The work output from the lower
cycle is

$$W_L = \eta_L Q_{HL}. \tag{1.12}$$

But

$$Q_{HL} = Q_B(1 - \eta_H), \tag{1.13}$$

where Q_B is the heat supplied to the upper plant, which delivers work

$$W_H = \eta_H Q_B. \tag{1.14}$$

Thus the total work output is

$$W = W_H + W_L = \eta_H Q_B + \eta_L(1 - \eta_H)Q_B$$
$$= Q_B(\eta_H + \eta_L - \eta_H\eta_L). \tag{1.15}$$

The thermal efficiency of the combined plant is therefore

$$\eta_{TH} = \frac{W}{Q_B} = \eta_H + \eta_L - \eta_H\eta_L. \tag{1.16}$$

The advantages are clear. The basic efficiency of the upper cycle has been
increased by $\eta_L(1 - \eta_H)$, by the use of its rejected heat to generate power
in the lower cycle.

An example of a combined cyclic power plant is one operating on
mercury and steam vapour cycles. The critical temperature of mercury is
about 1500°C, at a high critical pressure of 106 MPa, but at a heat supply

temperature of say 500°C its saturation pressure is quite low (less than 1 MPa). It can be condensed at modest temperature and pressure (avoiding the excessive vacuum required to attain ambient temperatures), rejecting heat to a bottoming steam cycle. However, technical problems associated with mercury have limited use of this combined cycle.

Less difficult from an engineering point of view is a combined power plant with a gas turbine rejecting heat from its exhaust to a closed-circuit steam turbine plant. The gas turbine may be closed-circuit or, more often, open circuit. There are several variants:

(i) fuel may be burnt in the gas turbine combustion chamber only, with heat supplied to the steam cycle from the gas turbine exhaust (and possibly also from the combustion chamber);

(ii) fuel may also be burnt in the gas turbine exhaust, where there is excess oxygen available for further combustion.

For the first case (with heat transfer from the exhaust only, Fig. 1.13) the work output from the gas turbine plant is

$$W_H = (\eta_o)_H F = (\eta_o)_H [(H_R)_0 - (H_P)_0], \qquad (1.17)$$

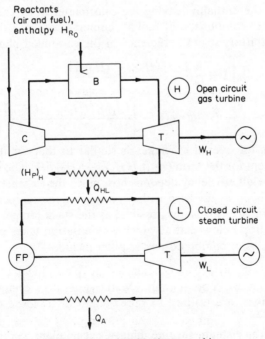

FIG. 1.13. Gas turbine/steam turbine
combined power plant.

where $(\eta_o)_H$ is the arbitrary overall efficiency defined in equation (1.6). The work output from the steam cycle is

$$W_L = (\eta_{TH})_L Q_{HL}, \tag{1.18}$$

where $(\eta_{TH})_L$ is the thermal efficiency of the lower cycle and Q_{HL} is the heat transferred from the gas-turbine exhaust.

Thus the arbitrary overall efficiency of the whole plant is

$$\eta_o = \frac{(W_H + W_L)}{F} = (\eta_o)_H + [(\eta_{TH})_L Q_{HL}/F]. \tag{1.19}$$

But if combustion is adiabatic, then the First Law of Thermodynamics for the open-circuit gas turbine (with exhaust of enthalpy $(H_P)_H$ leaving the plant) gives

$$(H_R)_0 = (H_P)_H + W_H + Q_{HL}, \tag{1.20}$$

so that

$$\begin{aligned}
Q_{HL} &= (H_R)_0 - (H_P)_H - W_H \\
&= [(H_R)_0 - (H_P)_0] + [(H_P)_0 - (H_P)_H] - W_H \\
&= F - [(H_P)_H - (H_P)_0] - W_H,
\end{aligned} \tag{1.21}$$

where $(H_P)_0$ is the enthalpy leaving the calorimeter in a "calorific value" experiment, after combustion of fuel at temperature T_0.

Hence the arbitrary overall efficiency of the combined plant is

$$\begin{aligned}
\eta_o &= (\eta_o)_H + (\eta_{TH})_L \left[1 - \frac{W_H}{F} - \frac{[(H_P)_H - (H_P)_0]}{F} \right] \\
&= (\eta_o)_H + (\eta_{TH})_L - (\eta_o)_H(\eta_{TH})_L - \left\{ \frac{(\eta_{TH})_L[(H_P)_H - (H_P)_0]}{F} \right\}.
\end{aligned} \tag{1.22}$$

This expression for overall efficiency is similar to that for the combined cycle plant, except for the term $(\eta_{TH})_L[(H_P)_H - (H_P)_0]/F$. The extent of this reduction in overall efficiency depends how much the exhaust gases can be cooled and is theoretically zero if they emerge at the temperature of the reactants. In practice this is not possible, as the stack temperature cannot be allowed to drop below that at which condensation takes place.

Other examples of combined power plant include

(i) a magneto-hydrodynamic/steam turbine plant, in which the exhaust gases from an "open-circuit" MHD generator, operating at high temperature, are utilised to raise steam for a closed-circuit steam plant;

(ii) a gas-turbine/steam-turbine binary cyclic plant in which heat is supplied to the gas turbine from a nuclear reactor.

Several combined power plants have operated successfully. Their adaptation to provide combined heat and power is a matter of continuing study and practical development, with a view to the maximum utilisation of fuel energy.

1.5 Reversibility, Availability and Exergy

The concepts of reversibility and irreversibility are important in the analysis of conventional power plant. Attainment of reversibility in the components of a power plant leads to greater overall plant efficiency, and the concept of thermodynamic availability may be used to demonstrate this. A brief summary of the results of availability theory for steady flow processes is given here (see also Appendix A).

(a) Reversible Flow in Presence of Environment at T_0

The maximum work which can be obtained in steady flow through a control volume (CV) between prescribed stable states (1 and 2), in the presence of a conceptual environment at temperature T_0, (i.e. with heat transfer to that environment only) is

$$[(W_{CV})_{REV}]_1^2 = B_1 - B_2, \tag{1.23}$$

where B is the steady flow availability function

$$B = H - T_0 S. \tag{1.24}$$

A corollary of this theorem is that the maximum work that can be extracted from fluid at prescribed state 1 is the *exergy*

$$E_1 = B_1 - B_0, \tag{1.25}$$

B_0 is the steady flow availability function at the so-called "dead" state, where the fluid is in thermodynamic equilibrium with the environment, at state (p_0, T_0). (A full description of the use of exergy in power plant analysis is given by Kotas[4].) The maximum work obtainable between states 1 and 2 may thus be written as

$$[(W_{CV})_{REV}]_1^2 = (B_1 - B_0) - (B_2 - B_0) = E_1 - E_2. \tag{1.26}$$

(b) Irreversible Flow in Presence of Environment at T_0

The actual work output in flow between states 1 and 2 is

$$[W_{CV}]_1^2 = [(W_{CV})_{REV}]_1^2 - T_0 \Delta S_{CV}, \tag{1.27}$$

where ΔS_{CV} is the entropy created within the control volume, so that $T_0 \Delta S_{CV}$ is the lost work due to irreversibility (sometimes simply called the irreversibility), $I = T_0 \Delta S_{CV}$.

(c)   Reversible Flow with Useful Heat Rejection at T_U

When useful heat $\int dQ_U$ is also rejected at variable temperature T_U in a steady-flow process between states 1 and 2 (but again in the presence of an environment at T_0), the (maximum) reversible work is

$$[(W_{CV})_{REV}]_1^2 = (B_1 - B_2) - \int_1^2 \left(\frac{T_U - T_0}{T_U}\right) dQ_U. \qquad (1.28)$$

(d)   Irreversible Flow with Useful Heat Rejection at T_U

Further, in a real (irreversible) process between states 1 and 2 the actual work output is

$$[W_{CV}]_1^2 = (B_1 - B_2) - \int_1^2 \left(\frac{T_U - T_0}{T_U}\right) dQ_U - I, \qquad (1.29)$$

where again $I = T_0 \Delta S_{CV}$ is the lost work due to irreversibility within the control volume, and ΔS_{CV} is the entropy creation within it—that part of the entropy change which cannot be accounted for by entropy fluxes associated with heat transfer.

(e)   Flows with Chemical Reaction

Application of equation (1.23) (or equation (1.25)) to a steady-flow process involving a chemical reaction, between reactants in a state $R_0(p_0, T_0)$ to products in a state $P_0(p_0, T_0)$ gives the maximum (reversible) work output as

$$(W_{CV})_{REV} = [(B_R)_0 - (B_P)_0] = [(G_R)_0 - (G_P)_0] = (-\Delta G_0), \quad (1.30)$$

where $G = H - TS$ in the Gibbs function. $(-\Delta G_0)$ is the quantity used in the definition of rational efficiency in equation (1.5(b)).

It follows (see Appendix A) that the difference between this maximum (reversible) work output $(-\Delta G_0)$ and the actual work output $[W_{CV}]_1^2$ from an open-circuit CHP plant (or a closed-circuit plant supplied with heat from an open-circuit boiler) from which products leave at state 2 with availability $B_2 = (B_P)_2$, is given by the equation

$$(-\Delta G_0) = [W_{CV}]_1^2 + \int_1^2 \left(\frac{T_U - T_0}{T_U}\right) dQ_U + T_0 \Delta S_{CV} + [(B_P)_2 - (G_P)_0] \quad (1.31)$$

Here it has been assumed that reactants are fed to the plant at the ambient condition so that $B_1 = (B_R)_1 = (G_R)_0$. The work potential lost in the real process thus comprises the quantities

(i) $\int_1^2 \left(\dfrac{T_U - T_0}{T_U}\right) dQ_U$, which is the work potential of the useful heat rejected in CHP plant;

(ii) $I = T_0 \Delta S_{CV}$, the irreversibility, and

(iii) $[(B_P)_2 - (G_P)_0]$, which is the exergy of the exhaust.

Equation (1.31) may be interpreted as an exergy flow equation (see Kotas[4]).

(f)   Loss of Work Output in Power Plants

Equations (1.29), (1.30) and (1.31) are useful in illustrating where work producing capacity is lost in the various processes of a power plant. The approach requires application of equation (1.29) to the individual components within the power plant. For example, in the adiabatic but irreversible flow through a turbine from state 1 to state 2 (see the temperature-entropy diagram of Fig. 1.14), the work output (per unit mass) will be the drop in specific enthalpy $[w_{CV}]_1^2 = (h_1 - h_2)$, if entry and exit velocities are small. The maximum work that could be achieved is $[(w_{CV})_{REV}]_1^2 = (b_1 - b_2)$. Thus the work lost due to irreversibility is

$$[(w_{CV})_{REV}]_1^2 - [w_{CV}]_1^2 = (b_1 - b_2) - (h_1 - h_2) = T_0(s_2 - s_1) = T_0 \Delta s,$$
(1.32)

the area A on Fig. 1.14.

An example of applying this type of analysis step by step through the various processes in a thermal power plant was undertaken by Traupel.[8] Figure 1.15 is taken from ref. 9 and illustrates the lost work and irreversibilities in unsupercharged and supercharged Diesel engines, of rational efficiency 40.1% and 43.8% respectively. The largest irreversibility occurs

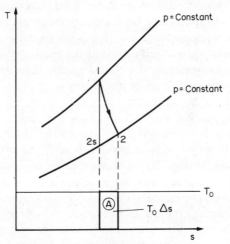

Temperature-entropy diagram

FIG. 1.14. Lost work in turbine expansion.

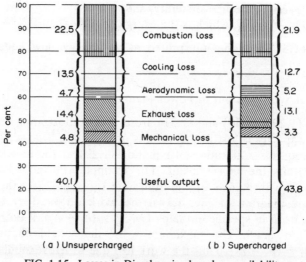

FIG. 1.15. Losses in Diesel engine based on availability
(Horlock[9], after Traupel[8]).

in combustion. The lost work due to aerodynamic irreversibility is shown, together with the lost work due to the heat rejection to cooling water, and that lost in the equivalent heat rejection in the high temperature exhaust.

Another study was undertaken by Horlock,[9] who split down the maximum possible work $(-\Delta G_0)$ into useful work, irreversibilities and lost work due to heat rejection—for both closed circuit and open circuit gas turbine plants. (The importance of this analysis was that the simple criterion of maximum thermal efficiency of the cyclic power plant may be misleading, for heat transfer to the closed circuit plant from an external heater may occur across a finite temperature drop, with substantial lost work resulting. Indeed the maximum overall efficiency may be approached by maximising the specific work of the closed circuit cycle of Fig. 1.3 rather than its thermal efficiency. The ratio of the thermal capacities of the external fluid flow (through control surface Z) and the internal fluid flow (through control surface Y) was shown to be critical.)

Similar analyses of power plant work output and "loss" of work have been comprehensively developed by Kotas,[4] who uses the concept of exergy flow. We return to the use of availability in Chapter 2, section 2.5, with particular reference to work output in the presence of useful heat rejection (as in CHP plant) and to the derivation of performance criteria for such plant.

References

1. Haywood, R. W. *Analysis of Engineering Cycles*, 3rd edition. Pergamon Press, Oxford, 1980.
2. Rogers, G. F. C. and Mayhew, Y. R. *Engineering Thermodynamics, Work and Heat Transfer*. Longmans, London, 1967.
3. Haywood, R. W. *Equilibrium Thermodynamics for Engineers and Scientists*. John Wiley, London, 1980.
4. Kotas, T. J. *The Exergy Method of Thermal Plant Analysis*. Butterworths, London, 1985.
5. Von Karman Institute for Fluid Dynamics. *Combined Cycles for Power Generation*. Lecture Series, Rhode Saint Genese, Belgium, 1978.
6. Benson, R. S. *Advanced Engineering Thermodynamics*, 2nd edition. Pergamon Press, Oxford, 1976.
7. Bammert, K. Twenty-five years of operating experience with the coal-fired closed cycle gas turbine cogeneration plant at Coburg. *Trans. ASME Journal of Engineering for Power*, **105**, 806–815, 1983.
8. Traupel, W. Reciprocating engine and turbine in internal combustion engineering. *Proc. CIMAC (Int. Cong. Combustion Engineers)* 37, 1957.
9. Horlock, J. H. The rational efficiency of power plants with external combustion. *Proc. Instn. Mech. Engrs.* **178**, 31, 43, 1963/64.

CHAPTER 2

Thermodynamics of CHP Plants—An Introductory Discussion (Some Examples, Performance Criteria, Effect of Irreversibility)

2.1 Introduction

The purpose of the designer of a combined heat and power (CHP) plant is different from that of the conventional power plant designer. Whereas the latter is attempting to meet a power demand (usually electrical) with maximum overall efficiency, the former is designing to meet both an (electrical) power demand *and* a heat load, by heat rejection from the CHP plant. Since both products have value, the overall efficiency is no longer an adequate criterion of performance.

In this chapter we shall review examples of CHP plant, develop thermodynamic criteria of performance, and give examples of their use.

2.2 Examples of CHP Plant

There are numerous examples of combined heat and power plant, most of which are developments or adaptations of the conventional power plant described in Chapter 1. A useful summary of some systems for

(a) supply of useful heat (Q_U),
(b) generation of (electrical) work (W), and
(c) supply of both useful heat (Q_U) and electrical work (W)

has been given by Timmermans.[1] Figure 2.1 shows representative values of Q_U and W, for a number of plants (examples A–K). The energy input of the fuel is unity ($F = 1.0$), so the overall efficiency $\eta_o = W$.

Domestic and industrial boilers (A and B) supply heat only, but the latter operate at considerably higher boiler efficiency (η_B). Conventional power plants (gas and steam turbines (C and D)), can attain overall

22

efficiencies of about 30% and 40% respectively. They may be linked in combined power plants to produce higher overall efficiency as explained in section 1.4.2; case E shows a combined power plant in which exhaust gases from the gas turbine raise steam for a closed cycle steam turbine, and Timmermans gives a representative overall efficiency of 44%.

The combined heat and power plants shown in Fig. 2.1 (F, G, H, J) include an extraction (or pass-out) steam power plant (case F) which for $Q_U = 0.1$ and $W = 0.38$. (But note that the useful heat (Q_U) may be substantially greater than this for large extraction rates, particularly for district heating schemes.) Case G shows that a steam power set operating with a back-pressure higher than in a conventional plant (to supply heat at the higher temperature required) may achieve values of $\eta_o = W = 0.25$ and $Q_U = 0.60$. Thermal efficiency is lost to produce the useful heat rejection. A gas turbine with waste heat recuperation (case H) may maintain its power output at $W = 0.30$, yet provide useful heat, $Q_U = 0.55$. Finally Timmermans shows a combined cycle with useful heat rejection (J). A gas turbine supplies its waste heat to a boiler for steam production to a back-pressure turbine. Some useful heat is also obtained from the gas turbine exhaust, which together with heat from the condensate of the back-pressure system provides a total useful heat $Q_U = 0.42$. Power output in the original gas turbine power output (0.30) plus steam turbine power output (0.1) gives a total $W = 0.40 = \eta_o$.

Most of the basic CHP systems are covered in Timmermans' list, but it is important to note that the gas turbine can be replaced by Diesel engines for relatively small power requirements. Another option for district heating which should be considered is that of a heat pump absorbing part of the power output from a conventional power plant (conventional power plant (example K)). The diagram shows a scheme proposed by Kolbusz,[2] in which a heat pump of high coefficient of performance ((CP = 6.6), but relatively low temperature lift, absorbs the work output (W) from a back-pressure turbine. Useful heat is thus supplied from the heat pump (at low temperature level) *and* from the condenser of the steam plant (at a higher temperature level). This is not strictly a CHP plant, as no net power is produced. The total heat output is substantially greater than the fuel energy input, because "free" heat has been abstracted from the environment.

Timmermans' figures are examples only and should not be taken as standards for the various types of plant shown, but they provide a useful introduction to CHP. There are many other more complex schemes, but the main CHP schemes are F (extraction condensing plant), G (back-pressure plant) and H (gas turbine or Diesel engine with waste heat recuperator).

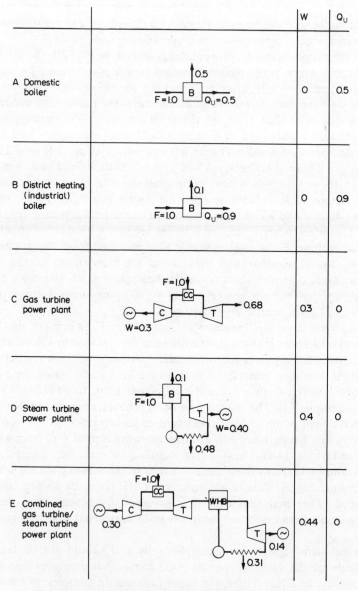

FIG. 2.1. Systems for separate generation of electricity (W) or heat (Q_U) showing energy flows (after Timmermans[1]).

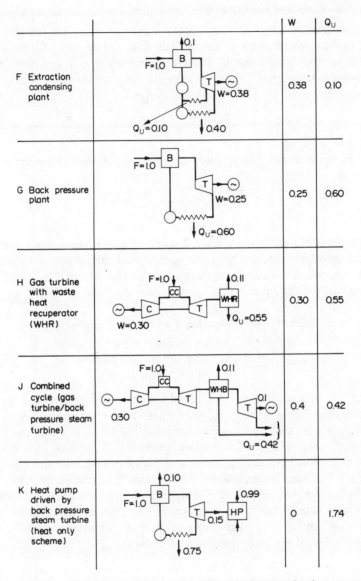

	W	Q_U
F Extraction condensing plant	0.38	0.10
G Back pressure plant	0.25	0.60
H Gas turbine with waste heat recuperator (WHR)	0.30	0.55
J Combined cycle (gas turbine/back pressure steam turbine)	0.4	0.42
K Heat pump driven by back pressure steam turbine (heat only scheme)	0	1.74

FIG. 2.1 (*cont.*) Systems for combined heat and power showing energy flows (after Timmermans[1] and Kolbusz[2]).

2.3 Performance Criteria for CHP Plants

2.3.1 *Efficiency and Energy Utilisation Factor*

As described in Chapter 1, the performance criterion for conventional closed circuit power plant is usually taken as the overall efficiency, the product of the closed cycle thermal efficiency (η_{TH}) and the boiler efficiency (η_B),

$$\eta_o = \eta_{TH}\eta_B = \frac{W}{F}, \tag{2.1a}$$

where $F = m_f(CV)_0$.

For an open circuit power plant an arbitrary overall efficiency is also defined as

$$\eta_o = \frac{W}{F}. \tag{2.1b}$$

These criteria of performance have less relevance to a combined heat and power plant which provides heat and generates electrical power. One more logical criterion is the energy utilisation factor (EUF), used by Porter and Mastanaiah[4] for example:

$$\text{EUF} = \frac{W + Q_U}{Q_B} \quad \text{(for a cyclic CHP engine),} \tag{2.2}$$

or
$$\text{EUF} = \frac{W + Q_U}{F} \quad \begin{array}{l}\text{(for the overall plant or an} \\ \text{open circuit plant),}\end{array} \tag{2.3}$$

where Q_U is the useful heat rejected which meets a required heat load, at a temperature T_U higher than T_0, the temperature of the environment. It is preferable not to use the term efficiency for the EUF, to avoid confusion with the thermal or overall efficiency. However, it must be remembered that work (W) is difficult to produce and highly priced, whereas the useful heat (Q_U) is a lower grade, lower priced product from the plant. As Polsky[3] and others have pointed out, the energy utilisation factor is thus therefore not entirely satisfactory as a criterion of performance as it gives equal weight to W and Q_U. Typical values of EUF for Timmermans' four CHP examples (F–J of Fig. 2.1) are given in Table 2.1.

2.3.2 *"Value-Weighted" Energy Utilisation Factor*

Another approach involves an attempt to account for the different pricing of electrical power and the heat load. If the sale value of electrical power is Y_E (£/kWh), that of the heat load is Y_H (£/kWh) and the price

TABLE 2.1. *Performance Criteria of CHP Plant (Timmermans' Examples of Fig. 1)*

Example	$(F)_{CG}$	$(W)_{CG}$	$(Q_U)_{CG}$	EUF	η_{eq}*	η_a†	FESR‡	$(IHR)_{CG}$§	λ_{CG}	(RC)
F Extraction condensing plant	1.0	0.38	0.10	0.48	0.41	0.43	0.057	2.33	0.26	0.41
G Back-pressure plant	1.0	0.25	0.60	0.85	0.45	0.75	0.235	1.33	2.40	0.40
H Gas turbine with waste heat recuperator	1.0	0.30	0.55	0.85	0.47	0.77	0.265	1.30	1.83	0.44
J Combined cycle with back-pressure system	1.0	0.40	0.42	0.82	0.54	0.75	0.318	1.33	1.05	0.50

* Useful heat weighted at a third of electricity value (Timmermans).
† Separate boiler efficiency $(\eta_B)_H$ taken as 0.9.
‡ Separate boiler efficiency $(\eta_B)_H$ taken as 0.9 and separate conventional power plant efficiency $(\eta_O)_C$ taken as 0.4.
§ Boiler or combustion efficiency $(\eta_B)_{CG} = 0.9$.

of fuel is $\mathscr{S}$ (£/kWh) then this "value-weighted" energy utilisation factor is

$$(\text{EUF})_{\text{vw}} = \frac{Y_E W + Y_H Q_U}{\mathscr{S} F} = \frac{Y_E}{\mathscr{S}}\left[\frac{W}{F} + \frac{Y_H}{Y_E}\frac{Q_U}{F}\right] = \left(\frac{Y_E}{\mathscr{S}}\right)\eta_{\text{eq}}, \quad (2.4)$$

where η_{eq} is referred to as an equivalent efficiency by Timmermans. Values of η_{eq} given by Timmermans are given in Table 2.1, for $(Y_H/Y_E) = 1/3$, i.e. with the value of useful heat taken as one third of that of electricity. This is a first, somewhat crude step towards taking economics into account.

2.3.3 Artificial Thermal Efficiency

An alternative criterion of performance sometimes used is an "artificial" thermal efficiency (η_a) in which the energy in the fuel supply to the CHP plant is supposed to be reduced by that which would be required to produce the heat load (Q_U) in a separate "heat only" boiler of efficiency $(\eta_B)_H$, i.e. by $(Q_U/(\eta_B)_H)$. The artificial efficiency (η_a) is then given by

$$\eta_a = \frac{W}{F - (Q_U/(\eta_B)_H)} = \frac{(\eta_o)_{\text{CG}}}{1 - \left(\dfrac{Q_U}{(\eta_B)_H F}\right)}, \quad (2.5)$$

where $(\eta_o)_{\text{CG}}$ is the overall efficiency of the CHP plant. Taking $(\eta_B)_H$ as 0.90, values of η_a have been calculated for Timmermans' four CHP examples and these are also given in Table 2.1.

2.3.4 Fuel Energy Savings Ratio

Another performance criterion developed for combined heat and power plant involves comparison between the fuel required to meet the given loads of electricity and heat in the CHP plant with that required in separate conventional plant to meet the same loads, say in a conventional electric power station of overall efficiency $(\eta_o)_C$ and a "heat only" boiler of efficiency $(\eta_B)_H$. Then the fuel energy saved is

$$\Delta F = \frac{Q_U}{(\eta_B)_H} + \frac{W}{(\eta_o)_C} - F,$$

and the fuel energy savings ratio (FESR) is defined as the ratio of the saving (ΔF) to the fuel energy required in the conventional plants

$$\text{FESR} = \frac{\Delta F}{\left(\dfrac{Q_U}{(\eta_B)_H} + \dfrac{W}{(\eta_o)_C}\right)}$$

$$= 1 - \frac{((\eta_o)_C/(\eta_o)_{\text{CG}}}{[1 + (\lambda_{\text{CG}}(\eta_o)_C/(\eta_B)_H)]}, \quad (2.6)$$

where $\lambda_{CG} = (Q_U/W)_{CG}$ is the (useful) heat to work ratio of the CHP plant, and $(\eta_o)_{CG}$ is its overall efficiency.

Again taking $(\eta_B)_H = 0.90$ and assuming $(\eta_o)_C = 0.40$ the fuel energy savings ratios Timmermans' four CHP examples have been calculated and are given in Table 2.1.

This thermodynamic criterion of performance is perhaps the most useful yet described as it can be used directly in economic assessment of CHP plant, as will be illustrated later in Chapters 3 and 4. Polsky,[3] who favours this approach, provides a useful discussion of how FESR varies with thermodynamic parameters (e.g. inlet and reheat conditions, back-pressure, etc.) in steam CHP schemes.

2.3.5 Incremental Heat Rate

Porter and Mastanaiah[4] introduce a thermodynamic criterion of performance called the incremental heat rate $(IHR)_{CG}$ in analysing the thermodynamics (and subsequently the economics) of CHP or cogeneration plant (subscript CG). We consider here the thermodynamic determination of $(IHR)_{CG}$, and return to their economic analysis subsequently in Chapter 4.

Consider a CHP plant delivering electrical power (W_{CG}) and meeting a (useful) heat load $((Q_U)_{CG})$ (Fig. 2.2). The fuel energy supplied is F_{CG}, either in a boiler plant of efficiency $(\eta_B)_{CG}$, or in the combustion chamber of an open-circuit plant (a gas turbine or a diesel engine) with a suitable efficiency for combustion, also taken as $(\eta_B)_{CG}$.

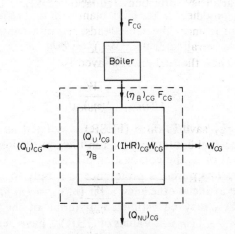

FIG. 2.2. Incremental heat rate
$(IHR)_{CG}$ for CHP plant.

From the first law of thermodynamics applied to the dotted control surface of Fig. 2.2

$$(\eta_B)_{CG} F_{CG} = (Q_U)_{CG} + W_{CG} + (Q_{NU})_{CG}, \tag{2.7}$$

where $(Q_{NU})_{CG}$ is the "non-useful" heat rejected.

The incremental heat rate is introduced from the determination of F_{CG}, which is arbitrarily assumed to be broken down as the sum of the fuel energy that *would* be supplied at the boiler (of combustion efficiency $(\eta_B)_{CG}$) to meet the heat load, and the balance which would go to the production of electricity (at an incremental heat rate $(IHR)_{CG}$) (see Fig. 2.2). Thus

$$F_{CG} = \frac{(Q_U)_{CG}}{(\eta_B)_{CG}} + (IHR)_{CG} W_{CG}, \tag{2.8}$$

so that from equations (2.7) and (2.8),

$$
\begin{aligned}
(IHR)_{CG} &= \frac{F_{CG}}{W_{CG}} - \frac{(Q_U)_{CG}}{(\eta_B)_{CG} W_{CG}} \\
&= \frac{1}{(\eta_o)_{CG}} - \frac{\lambda_{CG}}{(\eta_B)_{CG}}.
\end{aligned} \tag{2.9}
$$

If $(\eta_B)_{CG} \approx (\eta_B)_H$ then a simpler interpretation of $(IHR)_{CG}$ is that (by analogy with a conventional heat rate) it is the "artificial" heat rate of the cogeneration plant, i.e.

$$(IHR)_{CG} = \frac{1}{\eta_a},$$

where η_a is the "artificial" efficiency (equation (2.5)).

From equation (2.5)

$$
\begin{aligned}
\frac{1}{\eta_a} &= \frac{F_{CG} - ((Q_U)_{CG}/(\eta_B)_H}{W_{CG}} \\
&= \frac{1}{(\eta_o)_{CG}} - \left(\frac{\lambda_{CG}}{(\eta_B)_H}\right),
\end{aligned}
$$

which is equation (2.9), if $(\eta_B)_H = (\eta_B)_{CG}$.

Porter and Mastanaiah interpret $(\eta_B)_{CG}$ either as the boiler efficiency of a CHP steam plant or as the combustion efficiency of, say, an open cycle gas turbine plant. The author's preference is simply to take $(IHR)_{CG}$ as the inverse of the artificial efficiency η_a, if $(\eta_B)_{CG} \approx (\eta_B)_H$.

Like η_a, $(IHR)_{CG}$ may be estimated in terms of the thermodynamic parameters $(\eta_o)_{CG}$, λ_{CG}, $(\eta_B)_{CG}$. Values of $(IHR)_{CG}$ have been calculated for Timmermans' examples (for $(\eta_B)_{CG} = 0.9$) and are given in Table 2.1.

2.3.6   A Performance Criterion Based on Availability Analysis

In section 1.5 of Chapter 1, the application of availability (or exergy) analysis to conventional power plant was described. The basic equations given there (e.g. equation (1.29)) can be applied to CHP plant, for which the term

$$\int \left(\frac{T_U - T_0}{T_U} \right) dQ_U$$

becomes important. It represents the loss of work potential involved in the transfer of useful heat (Q_U) at temperature T_U. Kotas[5] has, for example, used these equations in studying exergy fluxes into and out of CHP plant.

The analysis of section 1.5 also enables a new "rational" criterion of performance for a cogeneration plant to be defined. Consider a CHP plant operating on an open circuit, producing work (W) and useful heat (Q_U) at a *constant* temperature (T_U); its "conventional" rational efficiency is

$$\eta_R = \frac{W}{(-\Delta G_0)}, \tag{2.10}$$

where $(-\Delta G_0) = (G_R)_0 - (G_P)_0$ is the change in Gibbs function that would occur if reactants entering at pressure p_0 and temperature T_0 changed to products at the same pressure temperature (see section 1.5).

But from equation (1.28) the maximum work output that could be achieved in flow through a control volume between stable states 1 and 2 is

$$[(W_{CV})_{REV}]_1^2 = (B_1 - B_2) - \int_1^2 \left(\frac{T_U - T_0}{T_U} \right) dQ_U, \tag{2.11}$$

if useful heat transfer takes place.

If the ratio of useful heat output to work output is fixed and equal to λ, then

$$\int_1^2 dQ_U = \lambda [(W_{CV})_{REV}]_1^2. \tag{2.12}$$

Further if T_U is constant, then equations (2.11) and (2.12) yield

$$[(W_{CV})_{REV}]_1^2 = \frac{(B_1 - B_2)}{1 + \lambda \left(1 - \dfrac{T_0}{T_U} \right)} \tag{2.13}$$

(see Horlock and Haywood[7]).

If the process between 1 and 2 is that involving combustion of reactants to products (i.e. $1 \equiv R_0$, $2 \equiv P_0$) then

$$[(W_{CV})_{REV}]_1^2 = \frac{(-\Delta G_0)}{1 + \lambda \left(1 - \dfrac{T_0}{T_U}\right)}, \tag{2.14}$$

and the maximum (work plus useful heat) is

$$[(W_{CV})_{REV}]_1^2 + (Q_U)_{REV} = \frac{(-\Delta G_0)(1 + \lambda)}{1 + \lambda \left(1 - \dfrac{T_0}{T_U}\right)}. \tag{2.15}$$

The *actual* output of (work plus useful heat) is

$$\begin{aligned} W + Q_U &= W(1 + \lambda) \\ &= \eta_R(-\Delta G_0)(1 + \lambda). \end{aligned} \tag{2.16}$$

So a rational criterion of CHP plant performance (RC) is

$$\begin{aligned} (RC) &= \frac{(W + Q_U)}{(W + Q_U)_{REV}} \\ &= \eta_R \left(1 + \lambda - \frac{\lambda T_0}{T_U}\right), \end{aligned} \tag{2.17}$$

which may be compared with the more conventional energy utilisation factor

$$EUF = \frac{W + Q_U}{(-\Delta H_0)} = (1 + \lambda)(\eta_0)_{CG}; \tag{2.18}$$

so that if $\eta_R \approx (\eta_0)_{CG}$

$$(RC) = EUF - \lambda(\eta_0)_{CG} \left(\frac{T_0}{T_U}\right). \tag{2.19}$$

Approximate values of (RC) are given in Table 2.1, assuming $(T_0/T_U) = 0.75$ for the four CHP examples F–J.

An alternative interpretation is that for a given heat demand (Q_U) and a given electrical work demand (W), the ratio of the fuel required in the ideal reversible case F_{REV} to that required in the real (irreversible) case (F) is

$$(RC) = \frac{F_{REV}}{F}. \tag{2.20}$$

In practical cases λ is often not so closely defined as a design parameter as implied here. An electrical load (W) may be met and as much useful heat sold as possible. Alternatively the CHP plant may be designed to meet a heat load (Q_U) and as much electricity sold as possible. However, the factor has validity even in those cases as a measure of thermodynamic performance, with the operational value of λ used in the above expressions.

2.4   Criteria for Component Performance

Conventional criteria of thermodynamic performance are used for many of the various components of a CHP plant, some of which are listed in Table 2.2. These criteria are well established and require little comment; ϵ and η_{cc} are not used subsequently in this book.

However, components not present in conventional power plant but important in CHP plant are the waste heat recuperator and the waste heat boiler, and some discussion of the criteria for assessment of their performance is required. Since steam is usually raised in these units they are sometimes referred to as heat recovery steam generators (unfired or fired).

The waste heat recuperator (WHR, or HRSG (unfired)) is essentially a heat exchanger, as it is not fired with additional liquid or gaseous fuel. Although the effectiveness (ϵ) as defined above is useful in the analysis of basic gas turbine cycle performance, it is less useful for a WHR as the heated fluid (cold side) is often water/steam, and the temperature rise (cold side) is not then significant. A more useful criterion is therefore an efficiency,

$$\eta_{WHR} = \frac{\text{Heat transferred to cold side}}{\text{Heat transferred from hot side}}$$

TABLE 2.2

Component	Criterion of performance		
Boiler	Efficiency,	$\eta_B = \dfrac{\text{Heat Output}}{\substack{\text{Enthalpy of combustion of fuel} \\ \text{at a standard temperature}}}$	(2.21)
Turbine	Isentropic efficiency,	$\eta_T = \dfrac{\text{Enthalpy drop}}{\text{Isentropic enthalpy drop}}$	(2.22)
Compressor	Isentropic efficiency,	$\eta_c = \dfrac{\text{Isentropic enthalpy rise}}{\text{Enthalpy rise}}$	(2.23)
Heat exchanger	Effectiveness (or thermal ratio)	$\epsilon = \dfrac{\text{Temperature rise (cold side)}}{\substack{\text{Maximum temperature difference between} \\ \text{entry (hot side) and entry (cold side)}}}$	(2.24)
Combustion chamber	Efficiency,	$\eta_{cc} = \dfrac{\substack{\text{Theoretical fuel–air ratio to produce a} \\ \text{given temperature}}}{\substack{\text{Actual fuel–air ratio needed to produce} \\ \text{the given temperature}}}$	(2.25)

Figure 2.3(a) shows a WHR receiving exhaust gas from a gas turbine with enthalpy $(H_G)_T$ at temperature T. Heat $(Q_U)_R$ is transferred to raise steam and the gases leave at a stack temperature $(T_G)_S$ with an enthalpy flux $(H_G)_S$. If there is a heat loss of L_R then

$$(H_G)_T = (H_G)_S + (Q_U)_R + L_R,$$

so that

$$\eta_{WHR} = \frac{(Q_U)_R}{[(H_G)_T - (H_G)_S]} = 1 - \frac{L_R}{[(H_G)_T - (H_G)_S]}. \qquad (2.26)$$

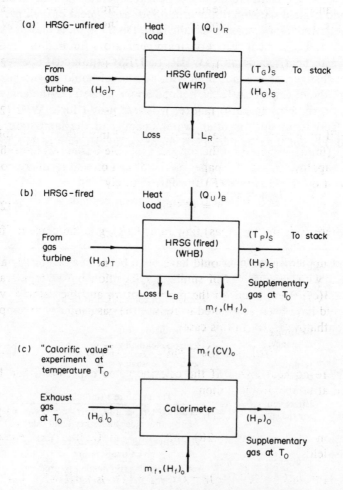

FIG. 2.3. Heat recovery steam generators (HRSG's).

It would appear to be logical to treat a waste heat boiler (WHB or HRSG (fired)), fed with additional liquid or gaseous fuel, as a conventional boiler with efficiency

$$\eta_B = \frac{\text{Heat transferred}}{\text{Calorific value of fuel supplied}}.$$

However (as in gas turbine combustion chambers) a difficulty arises in relation to specification of the temperature at which the calorific value is defined. Figure 2.3(b) shows a WHB receiving exhaust gas at the same temperature T (enthalpy $(H_G)_T$) as the WHR considered above. Supplementary gas (mass m_f) of enthalpy $(H_f)_0$ is supplied at a lower (near ambient) temperature T_0, and after combustion with the entering gas, heat $(Q_U)_B$ is transferred to raise steam and the gas products leave at a stack temperature $(T_P)_S$ with enthalpy flux $(H_P)_S$. If the heat loss is now L_B, then

$$(H_G)_T + (H_f)_0 = (Q_U)_B + (H_P)_S + L_B, \tag{2.27}$$

so that

$$(Q_U)_B = (H_G)_T + (H_f)_0 - (H_P)_S - L_B. \tag{2.28}$$

(We should note that there is a limit on the fuel input that is possible in an HRSG (fired), imposed by the oxygen available in the turbine exhaust gases. Ganapathy,[6] in a useful paper on the design of heat recovery boilers gives a limit on the heat rate $(\dot{F})$ as approximately

$$\dot{F} = 135.8\dot{m}_f g \tag{2.29}$$

in kW, where $\dot{m}_f$ is the fuel (gas) flow rate (kg/s), g is the percent "free" oxygen.)

But the supplementary gas could have been burnt in a "calorific value" experiment with exhaust gas of similar composition but at temperature T_0 (Fig. 2.3(c)). Heat equal to the product of m_f and the calorific value $(CV)_0$ would have been abstracted to reduce the gas products to temperature T_0 (enthalpy $(H_P)_0$). In this case

$$(H_G)_0 + (H_f)_0 = m_f(CV)_0 + (H_P)_0, \tag{2.30a}$$

if there were no heat losses. If the calorific value experiment had been conducted at temperature T, then

$$(H_G)_T + (H_f)_T = m_f(CV)_T + (H_P)_T. \tag{2.30b}$$

Combination of equations (2.28) and equation (2.30(a)) or equation (2.30(b)) yields

$$(Q_U)_B = m_f(CV)_0 + [(H_G)_T - (H_G)_0] - [(H_P)_S - (H_P)_0] - L_B, \tag{2.31a}$$

or

$$(Q_U)_B = m_f(CV)_T - [(H_f)_T - (H_f)_0] - [(H_P)_S - (H_P)_T] - L_B.$$
(2.31b)

To follow the conventional definition of boiler efficiency, we could use the second of these equations (2.31b), defining η_B as

$$\eta_B = \frac{(Q_U)_B}{m_f(CV)_T} = 1 - \frac{[(H_P)_S - (H_P)_T]}{m_f(CV)_T} - \frac{[(H_f)_T - (H_f)_0]}{m_f(CV)_T} - \frac{L_B}{m_f(CV)_T},$$
(2.32)

since it will always be less than unity. However, a more useful measure of performance for a WHB (for calculation of overall CHP plant performance) is from equation (2.31a)

$$\psi = \frac{(Q_U)_B}{m_f(CV)_0} = 1 + \frac{[(H_G)_T - (H_G)_0]}{m_f(CV)_0} - \frac{[(H_P)_S - (H_P)_0]}{m_f(CV)_0} - \frac{L_B}{m_f(CV)_0},$$
(2.33)

although ψ may be greater than unity, and will vary with the entry gas temperature T.

We can now relate the heat transferred in the recuperator (WHR) and the waste heat boiler. We may note that equation (2.28) for a WHR (HRSG, unfired) and equation (2.31(a)) for a WHB (HRSG, fired) may be combined to give

$$(Q_U)_B = (Q_U)_R + m_f(CV)_0 + \{L_R - L_B\}$$
$$+ \{[(H_G)_S - (H_G)_0] - [(H_P)_S - (H_P)_0]\}. \quad (2.34a)$$

If the two terms in curly brackets are each small then approximately

$$(Q_U)_B = (Q_U)_R + m_f(CV)_0, \quad (2.34b)$$

$$\psi = 1 + [(Q_U)_R/m_f(CV)_0]. \quad (2.34c)$$

Ganapathy[6] also gives a useful introduction to other aspects of HRSG design, including limitations imposed by the "pinch point"—the point in the boiler where water temperature approaches the hot gas temperature, just before evaporation.

2.5 The Effect of Internal Irreversibility on Performance of CHP Plant

The effects of internal irreversibility on the performance of conventional power plant are direct and important. Equation (1.21) showed the work output between states 1 and 2, in the absence of useful heat rejection

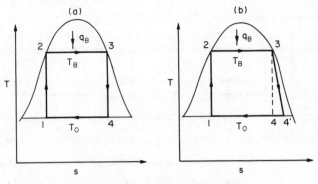

FIG. 2.4. Effect of irreversibility on Carnot steam power cycle.

(Q_U) at temperature T_U, to be reduced by $T_0 \Delta S_{CV}$, where ΔS_{CV} is the entropy creation internally. The effect on efficiency $(\eta_o)_C$ is usually in direct proportion.

But the effect of irreversibility on the performance of CHP plant requires further discussion. The presence of internal irreversibility, while reducing work output, may increase the useful heat rejection. To show the effect most simply, we consider first the basic Carnot power cycle, its modification for CHP and the effect of irreversibility on both cycles.

Figure 2.4(a) shows a basic steady-flow Carnot steam power cycle receiving heat in evaporation at constant pressure and temperature and rejecting heat in condensation at constant pressure and temperature. Figure 2.4(b) shows a modified steady-flow cycle in which the expansion takes place in an irreversible turbine. The work output from the Carnot cycle (per unit fluid circulating) is

$$w_{REV} = (h_3 - h_4) - (h_2 - h_1), \qquad (2.35)$$

and the work output from the irreversible cycle is

$$w = (h_3 - h_{4'}) - (h_2 - h_1) = \eta_T(h_3 - h_4) - (h_2 - h_1), \qquad (2.36)$$

where η_T is the turbine isentropic efficiency. The heat supply to each cycle is the same and equal to

$$q_B = (h_3 - h_2). \qquad (2.37)$$

The work output from the irreversible cycle is less than the maximum by

$$w_{REV} - w = (h_{4'} - h_4) = T_0(s_{4'} - s_4) = T_0 \Delta s, \qquad (2.38)$$

and the thermal efficiency is reduced,

$$\eta = \eta_{REV} - \frac{T_0 \Delta s}{q_B}. \qquad (2.39)$$

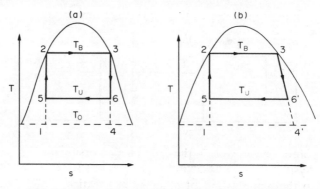

FIG. 2.5. Effect of irreversibility on Carnot (steam) CHP cycle.

Consider next a CHP or "back-pressure" Carnot cycle 5236 (Fig. 2.5(a)), and again suppose that the expansion turbine becomes irreversible (efficiency η_T) so that the cycle becomes (5236'), Fig. 2.5(b). The work output of the reversible cycle was

$$w_{REV} = (h_3 - h_6) - (h_2 - h_5). \tag{2.40}$$

The work output of the new (irreversible) cycle is

$$w = (h_3 - h_{6'}) - (h_2 - h_5), \tag{2.41}$$

so the work is reduced by

$$h_{6'} - h_6 = T_U(s_{6'} - s_6) = T_U \Delta s. \tag{2.42}$$

However, the useful heat rejection at temperature T_U is also changed. In the original reversible cycle it was

$$(q_U)_{REV} = (h_6 - h_5), \tag{2.43}$$

but in the irreversible cycle it is increased to

$$\begin{aligned} q_U &= (h_{6'} - h_5) \\ &= (h_{6'} - h_6) + (h_6 - h_5) \\ &= (q_U)_{REV} + T_U \Delta s. \end{aligned} \tag{2.44}$$

Clearly the thermal efficiency of the irreversible cycle is reduced compared with that of the reversible cycle ($\eta_{TH} < (\eta_{TH})_{REV}$), since $w < w_{REV}$ and the heat supply is unchanged. The energy utilisation factor of the reversible back-pressure Carnot cycle was unity,

$$(EUF)_{REV} = \frac{w_{REV} + (q_U)_{REV}}{q_B} = 1.0. \tag{2.45}$$

The energy utilisation factor of the irreversible new cycle is

$$EUF = (w + q_U)/q_B$$

$$= \frac{(w_{REV} - T_U\Delta s) + ((q_U)_{REV} + T_U\Delta s)}{q_B}$$

$$= (w_{REV} + (q_U)_{REV})/q_B$$

$$= (EUF)_{REV}, \tag{2.46}$$

and is therefore unchanged by the internal irreversibility.

This analysis of Carnot and "modified" Carnot (CHP) cycles is simply extended to the modification of other more practical cycles, such as the Rankine cycle. Figure 2.6 shows a reversible "back-pressure" Rankine cycle, and its modification for irreversibility if the turbine efficiency is η_T. Neglecting feed pump work, the work output (again per unit fluid circulating) is now

$$w = (h_3 - h_{6'}) = \eta_T(h_3 - h_6) = \eta_T w_{REV}, \tag{2.47}$$

and the new thermal efficiency is

$$\eta_{TH} = \eta_T(\eta_{TH})_{REV}. \tag{2.48}$$

The useful heat rejection in the reversible cycle was

$$(q_U)_{REV} = (h_6 - h_5), \tag{2.49}$$

but in the irreversible cycle it is

$$(q_U) = (h_{6'} - h_5), \tag{2.50}$$

and is increased by

$$q_U - (q_U)_{REV} = (h_{6'} - h_6) = T_U\Delta s. \tag{2.51}$$

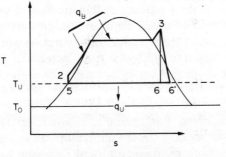

FIG. 2.6. Effect of irreversibility on
"back-pressure" Rankine cycle.

While the thermal efficiency is reduced by a factor η_T, the energy utilisation factor (EUF) is unchanged:

$$
\begin{aligned}
(\text{EUF}) &= \frac{w + q_U}{q_B} \\[2mm]
&= \frac{(w_{\text{REV}} - T_U \Delta s) + ((q_U)_{\text{REV}} + T_U \Delta s)}{q_B} \\[2mm]
&= \frac{w_{\text{REV}} + (q_U)_{\text{REV}}}{q_B} \\[2mm]
&= (\text{EUF})_{\text{REV}} \\[2mm]
&= 1.0.
\end{aligned}
\tag{2.52}
$$

Thus although the efficiencies of the irreversible cycles are less than the efficiency of the reversible cycles, their energy utilisation factors are unchanged. If the (EUF) were the criterion of performance the irreversible cycle would be as good as that of the reversible cycle; the attainment of reversibility (the objective in the thermodynamic analysis of conventional power cycles) would be open to question. However, we must remember that work production is difficult and costly so we should value provision of work more highly than the provision of useful heat. The use of availability theory in the analysis of CHP plant should not therefore be ignored; it remains a useful tool.

2.6   Simple Determination of Some Performance Parameters for CHP Plant

2.6.1   General

As indicated in Chapter 1, the objective in the design of conventional power plant is usually the achievement of maximum thermal efficiency— i.e. the production of the greatest possible amount of work from a given supply of fuel energy. However, even for conventional plant the facts of economic life may limit this objective. It may not be worthwhile investing a large amount of capital to produce a small increase in thermal efficiency; the economic return (the saving in fuel costs for a given power demand, or sales of more electricity for a given fuel supply) must be sufficient to meet the interest on the extra capital required, capital repayments and any extra maintenance costs. For example, analyses leading to the determination of the "economic" feed temperature and pressure in a steam power plant were given by Baumann[8] and discussed by Haywood[9] (but only take account of the interest on the extra capital required, not its repayment). Economic analyses for CHP plant are given later in this book, and are more comprehensive than the Baumann analysis for conventional plant.

The search for higher thermal efficiency is not so relevant in the study of CHP plant. Other performance parameters, such as the energy utilisation factor or the fuel energy savings ratio, prove more important. Further, the use of availability theory, important in conventional power plant when the maximum production of work is being sought, has to be re-examined for CHP plant.

As for conventional plant, economics as well as thermodynamics will have a part to play in the discussion of CHP plant performance. Since electrical power is more difficult to produce it will be more highly valued than the useful heat provided. There will therefore still be an emphasis on work production in CHP plant, but it has to be remembered that a second useful form of energy, which can be sold, is being produced. Timmermans[1] has reflected this point in his arbitrary definition of an equivalent efficiency

$$\eta_{eq} = \left[W + \left(\frac{Y_H}{Y_E} \right) Q_U \right]$$ (for unit fuel energy input) in his discussion of

power and CHP plant referred to in section 2.2.

Here we discuss the merits of two of the performance parameters given earlier in this chapter (thermal efficiency and energy utilisation factor) and the changes that take place in them as a power plant is modified for CHP. At this stage the discussion concentrates on thermodynamics, although the relative values of both electrical power and useful heat supplied are implicit in the discussion.

Again for simplicity, a first discussion can be based on the Carnot cycle. While it is recognised that this is not a practical cycle (although one which in practice it is sought to attain), the arguments for and against use of particular performance parameters are most easily demonstrated.

As described in the last section, a basic Carnot cyclic power plant (1234) (Fig. 2.4(a)) has a thermal efficiency $(\eta_{TH})_{CAR} = \left(\frac{T_B - T_A}{T_B} \right)$, where $T_2 = T_3 = T_B$ and $T_1 = T_4 = T_A$, and its energy utilisation factor is $(EUF)_{CAR} = (\eta_{TH})_{CAR}$ since all the heat rejection is "non-useful". The plant can be converted to a CHP plant by raising the back-pressure, and hence the rejection temperature from T_A to T_U (see Fig. 2.5(a)). The efficiency of this back pressure cogeneration plant drops to $(\eta_{TH})_{CG} = \left(\frac{T_B - T_U}{T_B} \right)$, but its energy utilisation is a maximum, with EUF = 1.0; no "non-useful" heat is rejected.

However, if the original Carnot power plant were converted to CHP (Fig. 2.7) by extraction of steam from the turbine at temperature T_U

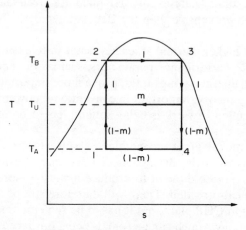

FIG. 2.7. Carnot cycle modified to
become extraction CHP plant.

(rather than by raising the back-pressure) then the thermal efficiency of
the plant would drop, and the energy utilisation factor would become less
than unity. If mass m were extracted along the turbine (for unit mass entering the high pressure turbine) then the new efficiency would be

$$(\eta_{TH})_{CG} = \frac{w}{q_B} = \frac{T_B - T_A(1 - m) - mT_U}{T_B}$$

$$= \left(\frac{T_B - T_A}{T_B}\right) - \frac{m(T_U - T_A)}{T_B}$$

$$= (\eta_{TH})_{CAR} - m\left(\frac{T_U - T_A}{T_B}\right). \qquad (2.53)$$

(If $T_U = T_A$, then $(\eta_{TH})_{CG} = (\eta_{TH})_{CAR}$, the original power plant efficiency.)
The energy utilisation factor of the CHP plant is

$$(EUF)_{CG} = \frac{q_U + w}{q_B}$$

$$= \frac{mT_U + (T_B - T_A) - m(T_U - T_A)}{T_B}$$

$$= \left(\frac{T_B - T_A}{T_B}\right) + \frac{mT_A}{T_B}$$

$$= (\eta_{TH})_{CAR} + \frac{mT_A}{T_B}. \qquad (2.54)$$

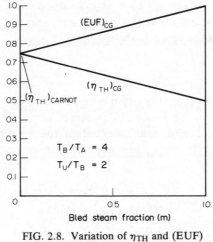

FIG. 2.8. Variation of η_{TH} and (EUF)
with bled steam fraction (m)
(modification of basic Carnot cycle,
T_B/T_A 4.0).

The energy utilisation factor (EUF)$_{CG}$ is less than unity, because non-useful heat is still being rejected at temperature T_A. (Clearly if all the steam is extracted at $T_U(m = 1)$ then the plant becomes a back-pressure set with efficiency $\eta_{TH} = \left(\dfrac{T_B - T_A}{T_B}\right)$ and (EUF)$_{CG} = 1.0$.)

By way of a numerical illustration, consider an example in which $\left(\dfrac{T_B}{T_A}\right) = 4$ and $\left(\dfrac{T_U}{T_A}\right) = 2$. Then $(\eta_{TH})_{CAR} = 0.75$, and

$$(\eta_{TH})_{CG} = 0.75 - (m/4), \qquad (2.55)$$

$$(EUF)_{CG} = 0.75 + (m/4). \qquad (2.56)$$

Figure 2.8 shows $(\eta_{TH})_{CG}$ and $(EUF)_{CG}$ plotted against m for this reversible CHP plant. For $m = 0$, the efficiency and the energy utilisation are the same, $(\eta_{TH})_{CG} = (EUF)_{CG} = (\eta_{TH})_{CAR} = 0.75$. As m is increased from zero so $(\eta_{TH})_{CG}$ drops from 0.75, and $(EUF)_{CG}$ increases above 0.75.

The case of $m = 1$ is the back-pressure turbine (all steam is abstracted at temperature T_U) described in the last section. For the numerical example given, the efficiency becomes 0.50 and the energy utilisation factor becomes unity.

This simple example using Carnot cycles serves to demonstrate that in a change from a power plant to a CHP plant thermal efficiency may be sacrificed to achieve higher energy utilisation. These are thermodynamic conclusions; whether it is worth making such a change will be subject to economic considerations, since *both* highly valued work and cheaper heat are available for sale (or to displace existing supplies). Similar analyses are now undertaken for more realistic plant.

2.6.2 The Z-Factor

For practical CHP plant it is often convenient to introduce a factor Z, defined as the ratio of the lost work (relevant to the conventional plant, subscript C) to the useful heat rejected in the modified (CHP) plant (subscript CG). The efficiency and energy utilisation factor of the modified (CHP) plant are then

$$(\eta_{TH})_{CG} = \frac{W_{CG}}{(Q_B)_{CG}} = \frac{W_C - Z(Q_U)_{CG}}{(Q_B)_{CG}}, \tag{2.57}$$

$$(EUF)_{CG} = \frac{(W_{CG} - Z(Q_U)_{CG}) + (Q_U)_{CG}}{(Q_B)_{CG}} = \frac{W_C + (1 - Z)(Q_U)_{CG}}{(Q_B)_{CG}}. \tag{2.58}$$

These equations again illustrate how energy utilisation may be gained at the expense of thermal efficiency in cogeneration plants.

Generally, for an extraction steam plant, as the bled steam flow is increased so the useful heat rejection is increased, but Z remains constant (i.e. the lost work is increased in proportion to the useful heat rejected). For example, for the "Carnot CHP" extraction plant of Fig. 2.7,

$$Z = \frac{m(T_U - T_A)}{mT_U} = \left(\frac{T_U - T_A}{T_U}\right), \tag{2.59}$$

and is constant, independent of m. Further, since the heat supplied is unchanged $(Q_B)_C = (Q_B)_{CG}$ it follows from (2.57) and (2.58) that

$$(\eta_{TH})_{CG} = (\eta_{TH})_{CAR} - m\left(\frac{T_U - T_A}{T_B}\right), \tag{2.60}$$

$$(EUF)_{CG} = (\eta_{TH})_{CAR} + m\left(\frac{T_A}{T_B}\right), \tag{2.61}$$

as derived before.

For a "Rankine CHP" extraction plant (Fig. 2.9),

$$Z = \frac{m(h_6 - h_4)}{m(h_6 - h_5)} = \frac{(h_6 - h_4)}{x_6 h_{fg}}, \tag{2.62}$$

where $(h_6 - h_4)$ is the enthalpy drop in the turbine beyond the extraction point, x_6 is the dryness fraction at state 6, and h_{fg} is the latent heat of

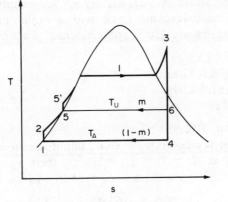

FIG. 2.9. Modification of Rankine
cycle for CHP, by steam extraction.

condensation at pressure $p_5 = p_6$. Again Z is independent of the bled
steam fraction m. Equations (2.57) and (2.58) are still valid, but the heat
supplied $(Q_B)_{CG}$ is not now the same as that in the original Rankine power
plant $(Q_B)_C$.

The thermal efficiency and the energy utilisation factor can be expressed
in terms of the bled steam fraction m, again neglecting feed pump work.
Since

$$(Q_U)_{CG} = mx_6 h_{fg}, \tag{2.63}$$

$$(Q_B)_{CG} = (h_3 - h_5) + (1 - m)(h_5 - h_2), \tag{2.64}$$

it follows that from (2.57), (2.58), (2.63) and (2.64) that

$$(\eta_{TH})_{CG} \approx \frac{(h_3 - h_4) - m(h_6 - h_4)}{(h_3 - h_5) + (1 - m)(h_5 - h_1)}, \tag{2.65}$$

$$(EUF)_{CG} \approx \frac{(h_3 - h_4) + \left[1 - \dfrac{(h_6 - h_4)}{x_6 h_{fg}}\right]mx_6 h_{fg}}{(h_3 - h_5) + (1 - m)(h_5 - h_1)}, \tag{2.66}$$

compared with the original value of thermal efficiency for the Rankine
power plant,

$$(\eta_{TH})_C \approx \frac{(h_3 - h_4)}{(h_3 - h_1)}. \tag{2.67}$$

The relevant enthalpies (h_1, h_5, h_3, h_6, h_4) are simply determined. For example, for a basic Rankine Cycle with a boiler pressure of 2 MPa, turbine inlet temperature 350°C and condenser pressure 7 kPa,

$$h_1 = 163.4 \text{ kJ/kg}$$
$$h_3 = 3138.6 \text{ kJ/kg}$$
$$h_4 = 2161.4 \text{ kJ/kg}$$

and $(\eta_{TH})_C = 0.3284$

(see Haywood,[9] p. 20).

The condenser temperature is 39°C, but if useful heat is rejected in the CHP plant at $T_5 = T_6 = T_U = 100°C$, then $h_5 = 417.5$ kJ/kg and $x_6 = 0.934$, $h_{fg} = 2257.9$ kJ/kg and $h_6 = 2526.4$ kJ/kg. It follows that

$$
(\eta_{TH})_{CG} = \frac{(3138.6 - 2161.4) - m(2526.4 - 2161.4)}{(3138.6 - 417.5) + (1 - m)(417.5 - 163.4)}
$$

$$
= \frac{0.3284 - 0.1227m}{1 - 0.0854m}, \tag{2.68}
$$

$$
(EUF)_{CG} = \frac{977.2 + \left[1 - \left(\dfrac{365.0}{0.934 \cdot 2257.9}\right)\right] 0.934 \cdot 2257.9}{2721.1 + (1 - m)254.1}
$$

$$
= \frac{0.3284 + 0.5861m}{1 - 0.0854m}. \tag{2.69}
$$

These equations again illustrate how energy utilisation may be gained at the expense of thermal efficiency in cogeneration plants.

Figure 2.10 shows the variation of $(\eta_{TH})_{CG}$ and $(EUF)_{CG}$ with the bled steam fraction, m. When $m = 0$, $(\eta_{TH})_{CG} = (EUF)_{CG} = 0.3284$, the basic Rankine cycle efficiency. When $m = 1.0$, the plant becomes a back-pressure set with a condensing temperature $T_U = 100°C$ and full energy utilisation.

2.6.3 Use of Availability Theory

These results may also be interpreted by employing availability theory, using equation (2.11) for the maximum work that could be obtained in steady flow through a control surface between states 2 and 3, for unit flow rate,

$$
[w_{REV}] = b_2 - b_3 - \int_2^3 \left(\frac{T_U - T_0}{T_U}\right) dq_U, \tag{2.11}
$$

(there is no creation of entropy $(\Delta s)_{CV}$ in the reversible cycles described in sections 2.6.1 and 2.6.2).

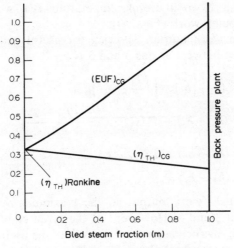

FIG. 2.10. Variation of η_{TH} and (EUF)
with bled steam fraction (m)
(modification of basic Rankine cycle).

Referring first to the Carnot cycle and its modifications for CHP (section 2.6.1 and Fig. 2.7), the term ($b_2 - b_3$) is the work output from the basic plant. For the CHP (extraction) plant the reduction in work output is

$$\int_2^3 \left(\frac{T_U - T_0}{T_U}\right) dq_U = m(T_U - T_0)(s_3 - s_2). \qquad (2.70)$$

If all the steam is extracted (a "back pressure" Carnot) then $m = 1$ and the reduction in work is

$$(T_U - T_0)(s_3 - s_2).$$

These reductions are reflected directly in the loss of cycle thermal efficiency, since the heat supplied between states 2 and 3 remains the same for these adaptations of the Carnot cycle to CHP.

Similar arguments can be applied to the more practical (but still reversible) Rankine power plant, if it is modified for CHP, either as an "extraction Rankine" with bled steam m, or with $m = 1$, as a "back-pressure Rankine". Equation (2.11) still applies, for it is a general expression for a reversible process. However, as explained in section 2.6.2, in these modifications to CHP plant the heat supplied is altered, whereas this was not the case for the modification of the Carnot power plant to CHP.

Consider first the simplest case of modification of the "conventional" Rankine cycle to a back-pressure set (by raising the heat rejection temperature from T_0 to a useful temperature T_U), with $m = 1.0$ and neglecting feed pump work (Fig. 2.9).

The maximum work output, allowing for useful heat rejection, that could be achieved between states 3 and 5 is

$$
\begin{aligned}
w_{REV} &= (b_3 - b_5) - \int \left(\frac{T_U - T_0}{T_U}\right) dq_U \\
&= (h_3 - h_5) - T_0(s_3 - s_5) - \left(\frac{T_U - T_0}{T_U}\right)(s_3 - s_5)T_U \\
&= (h_3 - h_5) - T_U(s_3 - s_5) \\
&= (h_3 - h_6)
\end{aligned}
\tag{2.71}
$$

This is the "actual" work output so the maximum reversible work is achieved between 3 and 5. However, for the given ambient temperature T_A, work $(h_6 - h_4)$ has been sacrificed to produce the useful heat rejection.

Similarly, for the original Rankine cycle modified to an "extraction-Rankine" plant, it may be shown that the work output is the maximum given by availability analysis, allowing for useful heat rejection at T_U but with no internal entropy creation (the cycle is still reversible).

2.7 Discussion

The first conclusion from this introductory study of the thermodynamics of CHP plant (and appropriate performance criteria) is that modification from a power plant to CHP increases the energy utilisation factor but at the expense of thermal efficiency. However, a more economic plant may be obtained, as both useful heat produced and the work output have (separate) values. The thermal efficiency is no longer a sufficient criterion of performance. The energy utilisation factor is more useful, but has its limitations since the value of useful heat supply is less than the value of electrical power output. We shall see later that the fuel energy savings ratio has more significance, since it has more direct relevance to the subsequent economic analyses which are required.

Secondly, as in the thermodynamics of conventional plant, the concept of availability has relevance since it tells us the maximum work that can be obtained allowing for useful heat rejection. Lost work due to internal irreversibility may increase the useful heat rejection in a CHP plant, but since work is more highly valued this may not be advantageous economically. Availability theory is therefore still useful in helping to assess the maximum work that can be obtained.

References

1. Timmermans, A. R. J. Combined Cycles and their Possibilities Lecture Series, Combined Cycles for Power Generation. Von Karman Institute for Fluid Dynamics, Rhode Saint Genese, Belgium, 1978.
2. Kolbusz, P. The Use of Heat Pumping in District Heating Systems. Electricity Research Council Report No. ECRC/M 700, 1974.
3. Polsky, M. P. Fuel Effectiveness of Cogeneration. *A.S.M.E. Paper 80*, JPGC/Pwr-8, 1980.
4. Porter, R. W. and Mastanaiah, K. Thermal-Economic Analysis of Heat-Matched Industrial Cogeneration Systems, *Energy*, **7**, 2, 171–187, 1982.
5. Kotas, T. J. *The Exergy Method of Thermal Plant Analysis*, Butterworths, London, 1985.
6. Ganapathy, V. Heat Recovery Boiler Design for Cogeneration, *Oil and Gas Journal*, Technology, 116–125, December, 1985.
7. Horlock, J. H. and Haywood, R. W. Thermodynamic Availability and its Application to Combined Heat and Power Plant, *Proc. Instn. Mech. Engrs.*, **199**, C1, 11–17, 1985.
8. Baumann, K. Some Considerations Affecting the Future Development of the Steam Cycle, *Proc. Inst. Mech. Engrs.*, **155**, 125, 1946.
9. Haywood, R. W. *Analysis of Engineering Cycles* (3rd Edition), Pergamon Press, Oxford, 1980.

CHAPTER 3

Comparative Thermodynamic Performance of CHP Plants (Fuel Savings, Choice of Plant)

3.1 Introduction

A brief outline of some CHP schemes has been given in Chapter 2. Performance criteria were defined, and estimates of such criteria made for some reversible CHP plant. Here we undertake more general assessments of the criteria of thermodynamic performance—overall efficiency (η_o), energy utilisation factor (EUF) and fuel energy savings ratio (FESR)—for several cogeneration plants, in terms of the performance of the separate units making up the combined plants, and the ratio of heat to work loads. These analyses lead to choice of the best type of plant for a given demand (but at this stage using thermodynamic criteria only).

It is important to realise what boundary conditions are imposed on the designer of a CHP plant. A critical issue is whether or not the plant is self-contained (i.e. separate from the central or utility grid). If the plant is not self-contained, then surplus capacity may be used elsewhere. The heat load may be met from the CHP plant and surplus electricity exported to the grid; alternatively, where an electrical demand exceeds that which the plant can provide it may be possible to import electricity from the grid—at a price. If the electrical demand is met and the heat load exceeds that which the plant can supply it may be possible to use stand-by boilers to meet the excess. Clearly the overall economic performance will not only depend on the thermodynamic performance, but also critically upon the other factors, such as the sale price of electricity or the "stand-by" cost of electricity from the grid, and the sale price of heat or the cost of heat from standby boilers.

In this chapter we consider the thermodynamics of CHP plants in the following way.

First, in section 3.2 we discuss the performance, and in particular the fuel consumption, of two separate plants (conventional electrical power plant and conventional heat only boiler plant) meeting prescribed electricity and heat loads; this is the reference case.

Next, in section 3.3 we present a simple analysis of the "perfectly matched" CHP plant, exactly meeting the heat and electricity demands (and their ratio) and assess its performance through criteria such as EUF and FESR.

In section 3.4 we consider the overall performance of "unmatched" plants, where the ratio of heat to electricity developed by the plant (λ_{CG}) is not the same as the demand value (λ_D). Either the CHP plant can meet the heat load, and electricity can be imported from or exported to the grid (usually the case with steam turbine plant—examples F and G of Table 2.1); or the CHP plant may meet the electrical load, and extra heat is provided either by a supplementary "heat only" boiler or by burning more fuel in the exhaust from a gas turbine (example H of Table 2.1). The case of a heat pump, driven either from the grid or from a back-pressure turbine (example K of Table 2.1) is also considered. These are essentially "design point" analyses at this stage; it is the performance, and particularly the fuel consumption, of the *whole* plant (CHP plant plus grid turbine, or CHP plant plus supplementary boiler) that has to be assessed, in comparison with the reference case.

"Carnot" CHP plants are analysed in section 3.5. Various practical plants are considered in section 3.6 and numerical examples are given. It is clear that heat to power ratios (both demand λ_D and the ratio the CHP plant can provide (λ_{CG})) have major effects on overall performance. In these calculations, and in the earlier analysis, it is assumed that the performance parameters for the CHP plant (efficiency ($\eta_o)_{CG}$, heat pump coefficient of performance (CP) etc.) are known or can be determined from the usual cycle analysis.

However, in section 3.7 we report work by Porter and Mastanaiah[1] which gives typical values of λ_{CG} and associated performance parameters (e.g. FESR and incremental heat rate (IHR)), for CHP plants operating alone, as "self-contained" units.

In section 3.8 we consider the scope for varying the ratio of heat to power produced by the plant, within its (thermodynamic) design, and how performance criteria change as a result. A particular study is made of the steam extraction CHP plant.

In all these analyses it is the fuel saving (in comparison with a "reference" plant) that is the critical factor and the FESR is the associated performance criterion. In section 3.9 we therefore consider a realistic example in which the fuel saving resulting from installation of a CHP plant in a district heating scheme is estimated.

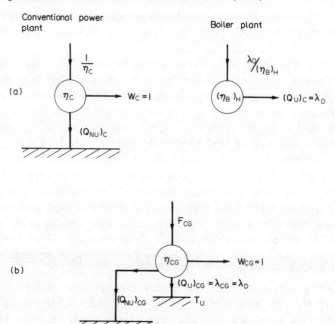

FIG. 3.1. (a) Reference system. Separate plants meeting
power demand of unity and heat load of λ_D.
(b) Replacement CHP plant, matched to produce power
demand of unity and heat load of λ_D.

3.2 Performance of Separate Conventional Plants—the Reference Case

Figure 3.1(a) shows diagrammatically a conventional electric power plant meeting the electricity load, and a conventional boiler meeting the heat load. This is the basic case which any CHP plant has to better in performance. The efficiency of the conventional electric power plant is $\eta_C{}^*$, and the electrical load is unity (it is assumed that the plant can be scaled to meet the actual loads). The ratio of heat to electrical demands is λ_D, so that heat load is taken as λ_D. The efficiency of the "heat only" boiler is $(\eta_B)_H$ (or η_B)* so the fuel energy required for the boiler is (λ_D/η_B), i.e. there are heat losses $\lambda_D\left(\dfrac{1}{\eta_B} - 1\right)$ involved before heat is delivered

---

*From here onwards $(\eta_o)_C$ is written as η_C—i.e. η implies overall efficiency, unless otherwise stated; also $(\eta_B)_H$ is now written as η_B. The efficiency $(\eta_B)_{CG}$ will not usually appear separately as it is included within the definition of η_{CG} ($= (\eta_B)_{CG}(\eta_{TH})_{CG}$).

to district or process heating. The total fuel energy required in this reference case is therefore

$$F_{REF} = \frac{1}{\eta_C} + \frac{\lambda_D}{\eta_B}$$

$$= \frac{\eta_B + \lambda_D \eta_C}{\eta_C \eta_B} \tag{3.1a}$$

$$= \frac{1 + \lambda_D \eta_C}{\eta_C}, \quad \text{if} \quad \eta_B = 1.0. \tag{3.1b}$$

The overall thermal efficiency of the two plants in producing electrical work is

$$\eta_{REF} = \frac{1}{F_{REF}}$$

$$= \frac{\eta_C \eta_B}{\eta_B + \lambda_D \eta_C}, \tag{3.2}$$

but this is not a significant criterion. More important is the energy utilisation factor

$$(EUF)_{REF} = \frac{(1 + \lambda_D)}{F_{REF}} = \frac{(1 + \lambda_D)\eta_C \eta_B}{(\eta_B + \eta_C \lambda_D)}; \tag{3.3}$$

if $\eta_B = 1.0$ then

$$(EUF)_{REF} = \frac{(1 + \lambda_D)\eta_C}{(1 + \eta_C \lambda_D)}. \tag{3.4}$$

This reference case will be used as a comparison in subsequent analyses, in particular to enable the fuel energy savings ratio (FESR) of a CHP plant to be determined.

3.3 The "Perfectly Matched" CHP Plant

If the replacement CHP plant (Fig. 3.1(b)) is exactly matched to the power and heat demands (i.e. the ratio λ_{CG} achieved by the CHP plant alone is equal to the demand value λ_D) then the fuel saving involved in the substitution is easy to determine. The fuel supply to the "self-contained" CHP plant is simply

$$F_{CG} = \frac{W}{\eta_{CG}} = \frac{1}{\eta_{CG}}. \tag{3.5}$$

Although the plant is matched to the power and useful heat demands, it may still reject "non-useful" heat $(Q_{NU})_{CG}$ (at a lower temperature than the "useful" temperature T_U) so that

$$\eta_{CG} = \frac{1}{F_{CG}}$$

$$= \frac{1}{1 + \lambda_D + (Q_{NU})_{CG}}. \qquad (3.6)$$

The energy utilisation factor is

$$(EUF)_{CG} = \frac{(1 + \lambda_D)}{F_{CG}} = (1 + \lambda_D)\eta_{CG} = \frac{(1 + \lambda_D)}{(1 + \lambda_D) + (Q_{NU})_{CG}}, \qquad (3.7)$$

which in general is less than unity.

The fuel savings, compared with the reference plant are

$$F_{REF} - F_{CG} = \frac{\lambda_D}{\eta_B} - \left(\frac{1}{\eta_{CG}} - \frac{1}{\eta_C}\right), \qquad (3.8)$$

and the savings produced by elimination of the separate boiler plant are reduced because of the lower overall efficiency of the "matched" cogeneration plant.

The fuel savings ratio is

$$FESR = \frac{F_{REF} - F_{CG}}{F_{REF}} = 1 - \frac{\eta_C \eta_B}{\eta_{CG}(\eta_B + \lambda_D \eta_C)}, \qquad (3.9)$$

where $\lambda_D = \lambda_{CG}$ in this case.

3.4 Overall Performance of "Unmatched" CHP Plants

We consider next the performance of the more common forms of CHP plant. We now make no initial assumption that the plants are perfectly matched to the heat and power demands $(\lambda_{CG} \neq \lambda_D)$. In selecting the type of plant, we are implying an associated value of λ_{CG} (although the value can vary somewhat according to design parameters—see section 3.6). We then consider how this CHP plant performs in meeting a demand requirement of λ_D, by operating in parallel with a conventional power plant or with a supplementary heat supply.

3.4.1 Back-Pressure Steam Turbine

Figure 3.2 shows diagrammatically a back-pressure steam turbine (example F of Fig. 2.1). We assume initially that although it is able to meet the required heat load, it is not matched to the power load, so that a con-

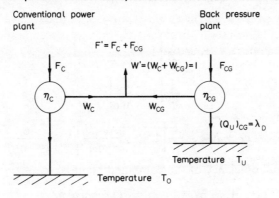

FIG. 3.2. Conventional and back pressure
plants operating in parallel to meet loads
$(1, \lambda_D)$.

ventional plant has to be operated in parallel, meeting some of the electrical load (i.e. the overall system absorbs electrical power from the grid). Here, and in the subsequent examples, subscript CG refers to the CHP plant alone, subscript C to the conventional plant operating in parallel, and superscript (′) refers to the overall performance of the total system replacing the reference plant. If under certain circumstances the conventional plant (C) is not needed then the superscript (′) becomes identical to the subscript (CG).

The two plants together produce a work output equal to unity. They require fuel energies F_C (conventional plant) and F_{CG} (back-pressure set). The conventional plant, of efficiency η_C, produces electrical power W_C and rejects heat to a reservoir at low (ambient) temperature. The back-pressure turbine, of overall efficiency η_{CG}, produces power W_{CG} and supplies useful heat to a higher temperature reservoir for process or district heating; it rejects $((Q_U)_{CG})$ but no non-useful heat $((Q_{NU})_{CG} = 0)$. It follows that

$$\eta_C F_C = W_C, \tag{3.10}$$

$$\eta_{CG} F_{CG} = W_{CG}, \tag{3.11}$$

$$(1 - \eta_{CG}) F_{CG} = \lambda_D, \tag{3.12}$$

since $\qquad \lambda_D = (Q_U)_{CG}/(W_{CG} + W_C) = (Q_U)_{CG}/1.$

For the two plants together

$$W' = W_C + W_{CG} = 1 = \eta_C F_C + \eta_{CG} F_{CG}$$

$$= \eta_C F_C + \frac{\eta_{CG} \lambda_D}{(1 - \eta_{CG})}, \tag{3.13}$$

so that

$$F_C = \frac{1}{\eta_C}\left[1 - \frac{\eta_{CG}\lambda_D}{(1 - \eta_{CG})}\right], \tag{3.14a}$$

$$= \frac{1}{\eta_C}\left[1 - \frac{\lambda_D}{\lambda_{CG}}\right]; \tag{3.14b}$$

$$F' = F_C + F_{CG} = \frac{1}{\eta_C}\left[1 - \frac{\eta_{CG}\lambda_D}{(1 - \eta_{CG})}\right] + \frac{\lambda_D}{(1 - \eta_{CG})}, \tag{3.15}$$

and the overall efficiency of the combined plant is

$$\eta' = \frac{W_C + W_{CG}}{F'} = \frac{(1 - \eta_{CG})\eta_C}{(1 - \eta_{CG}) + \lambda_D(\eta_C - \eta_{CG})}. \tag{3.16}$$

The energy utilisation factor is

$$(EUF)' = \frac{(1 + \lambda_D)}{F'} = \frac{(1 + \lambda_D)(1 - \eta_{CG})\eta_C}{(1 - \eta_{CG}) + \lambda_D(\eta_C - \eta_{CG})}. \tag{3.17}$$

The fuel energy savings ratio is

$$(FESR)' = \frac{F_{REF} - F'}{F_{REF}} = 1 - \eta_B\left[\frac{(1 - \eta_{CG}) + \lambda_D(\eta_C - \eta_{CG})}{(1 - \eta_{CG})(\eta_B + \eta_C\lambda_D)}\right]. \tag{3.18}$$

Several points of interest follow from these expressions, for the case of the back-pressure turbine matching the heat load exactly. We consider variations of λ_D (fixing the heat load but varying the power demand).

(i) As $\lambda_D \to \left(\dfrac{1 - \eta_{CG}}{\eta_{CG}}\right) = \lambda_{CG}$, $F_C \to 0$, $\eta' \to \eta_{CG}$, $(EUF)' \to 1.0$
   (i.e. the back-pressure plant is used alone).

As λ_D is increased (power demand decreased), the grid is used less; when $\lambda_D = \lambda_{CG}$ it is not used at all. This is because the back-pressure turbine is then exactly matched to the ratio of heat load to electric load; equations (3.16) to (3.18) become identical with equations (3.6), (3.7) and (3.9), with $(Q_{NU})_{CG} = 0$.

(ii) For λ_D exceeding λ_{CG}, equation (3.14) suggests that the fuel energy supplied to the grid turbine (F_C) becomes negative. With heat load held constant, the power demand has been decreased so much that the back-pressure steam turbine (or gas turbine) can now supply excess power *to* the grid. The work output from the conventional plant,

$$W_C = \left(1 - \frac{\lambda_D}{\lambda_{CG}}\right),$$

becomes negative when $\lambda_D > \lambda_{CG}$, so that

$$|W_C| = \left(\frac{\lambda_D}{\lambda_{CG}} - 1\right)$$

is exported to the grid and fuel energy

$$F_C = \frac{1}{\eta_C}\left[\frac{\lambda_D}{\lambda_{CG}} - 1\right]$$

is saved elsewhere in the grid. (Equation (3.14b) applies only for $\lambda_D \leq \lambda_{CG}$.)

A simple modification may be made to allow for heat loss between the back-pressure condenser and the district heating load (λ_D). If this heat loss is $\phi\lambda_D$, then the heat from the back-pressure turbine must be $(1 + \phi)\lambda_D$, so that

$$\eta' = \frac{(1 - \eta_{CG})\eta_C}{(1 - \eta_{CG}) + (1 + \phi)\lambda_D(\eta_C - \eta_{CG})}, \qquad (3.19)$$

$$(EUF)' = \frac{(1 + \lambda_D)}{F'}$$

$$= \frac{(1 + \lambda_D)(1 - \eta_{CG})\eta_C}{(1 - \eta_{CG}) + (1 + \phi)\lambda_D(\eta_C - \eta_{CG})}, \qquad (3.20)$$

$$(FESR)' = 1 - \eta_B\left[\frac{(1 - \eta_{CG}) + (1 + \phi)\lambda_D(\eta_C - \eta_{CG})}{(1 - \eta_{CG})(\eta_B + \eta_C\lambda_D)}\right]. \qquad (3.21)$$

We next consider the case of the back-pressure turbine only partly meeting the heat load, but fully meeting the electrical load. Figure 3.3 shows the heat load supplied partly by heat rejection from the back-pressure turbine (of efficiency η_{CG}) and partly by a supplementary boiler of efficiency η_B receiving fuel energy F_B. The power plants supplying the main grid are now not used—it being implied that $\lambda_D > \lambda_{CG}$; the CHP plant is scaled to meet the total electricity load and a heat only boiler introduced. It is assumed that this supplementary "heat only" boiler has the same efficiency as the "heat only" boiler of the reference case (η_B). It follows that

$$W_{CG} = 1 = \eta_{CG}F_{CG}, \qquad (3.22)$$

$$(Q_U)_{CG} = \frac{1 - \eta_{CG}}{\eta_{CG}}, \qquad (3.23)$$

$$(Q_U)' = \lambda_D = (Q_U)_{CG} + \eta_BF_B; \qquad (3.24)$$

so that $\qquad \left(\dfrac{1 - \eta_{CG}}{\eta_{CG}}\right) + \eta_BF_B = \lambda_D;$

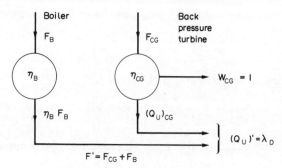

FIG. 3.3. Back pressure turbine with
supplementary heat only boiler.

and
$$F_B = \frac{1}{\eta_B}\left[\lambda_D - \frac{1 - \eta_{CG}}{\eta_{CG}}\right] = \frac{1}{\eta_B}(\lambda_D - \lambda_{CG}). \tag{3.25}$$

The last equation shows that F_B is positive only when $\lambda_D > \lambda_{CG}$ $\left(= \dfrac{1 - \eta_{CG}}{\eta_{CG}}\right)$, as expected from the earlier analysis. When $\lambda_D = \lambda_{CG}$, no supplementary boiler is required and the basic plant operates as a "matched" back-pressure set.

The overall efficiency is

$$\eta' = \frac{1}{F'} = \frac{\eta_B \eta_{CG}}{[\eta_B + \lambda_D \eta_{CG} + \eta_{CG} - 1]}$$

$$= \frac{\eta_{CG} \eta_B}{[\eta_B - \eta_{CG}(\lambda_{CG} - \lambda_D)]}. \tag{3.26}$$

The energy utilisation factor is

$$(\text{EUF})' = \frac{(\lambda_D + 1)\eta_{CG}\eta_B}{[\eta_B - \eta_{CG}(\lambda_{CG} - \lambda_D)]}, \tag{3.27}$$

and the fuel energy savings ratio is

$$(\text{FESR})' = 1 - (F'/F_{\text{REF}})$$

$$= 1 - \frac{\eta_C[\eta_B - \eta_{CG}(\lambda_{CG} - \lambda_D)]}{\eta_{CG}(\eta_B + \eta_C \lambda_D)}. \tag{3.28}$$

If η_B is unity then $(\text{FESR})'$ becomes

$$(\text{FESR})' = \frac{(1 - \eta_C)}{(1 + \eta_C \lambda_D)}, \tag{3.29}$$

which will be the highest (FESR) that can be achieved.

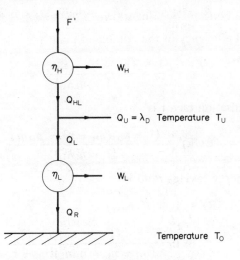

FIG. 3.4. Extraction steam turbine
plant, interpreted as a combined CHP
plant.

3.4.2   *Pass-Out or Extraction Turbine*

We consider next a pass-out or extraction turbine (example F of Fig.
2.1). Figure 3.4 shows an interpretation of how an extraction turbine can
perform as a CHP plant. We may regard it at this stage as a combined
"replacement" plant, the upper part operating at efficiency η_H, producing
work W_H from fuel energy F', and rejecting heat Q_{HL}. This heat meets the
heat load $(Q_U = \lambda_D)$ through condensation of the bled steam, and also
(through condensation and subsequent evaporation) effectively supplies
heat Q_L to the lower part of the cycle, which may be regarded as producing
work W_L at efficiency η_L.

It follows that

$$\eta_H F' = W_H, \tag{3.30}$$

$$(1 - \eta_H)F' = Q_{HL}, \tag{3.31}$$

$$\eta_L Q_L = W_L, \tag{3.32}$$

$$Q_{HL} = Q_L + Q_U = Q_L + \lambda_D = Q_L + \lambda_{CG} \tag{3.33}$$

The total work output is again unity,

$$W = W_H + W_L = 1,$$

so that $\lambda_D = Q_U/(W_H + W_L) = Q_U/1$.

But
$$\begin{aligned} W &= \eta_H F' + \eta_L Q_L \\ &= \eta_H F' + \eta_L[F'(1 - \eta_H) - \lambda_D], \end{aligned} \tag{3.34}$$

so that $$F' = (1 + \eta_L \lambda_D)/(\eta_H + \eta_L - \eta_H \eta_L), \tag{3.35}$$

and the overall efficiency of the combined plant is

$$\eta' = \frac{1}{F'} = \frac{(\eta_H + \eta_L - \eta_H \eta_L)}{(1 + \eta_L \lambda_D)}. \tag{3.36}$$

The energy utilisation factor is

$$(EUF)' = \frac{(1 + \lambda_D)(\eta_H + \eta_L - \eta_H \eta_L)}{(1 + \eta_L \lambda_D)}, \tag{3.37}$$

and the fuel energy savings ratio is

$$(FESR)' = 1 - (F'/F_{REF})$$

$$= 1 - \frac{(1 + \eta_L \lambda_D)(\eta_C \eta_B)}{(\eta_H + \eta_L - \eta_H \eta_L)(\eta_B + \eta_C \lambda_D)}. \tag{3.38}$$

A range of values of λ_D can be obtained by varying the amount of bled steam (but scaling the plant to meet the power demand). Points of interest are as follows.

(i) If $\lambda_D = 0$, then $\eta' = (EUF)' = \eta_H + \eta_L - \eta_H \eta_L$. This is the well-known expression for the overall efficiency of a combined power plant, in which the heat rejected from the upper cycle becomes the heat supplied to the lower cycle (see Chapter 1, equation (1.16)).

(ii) The heat supplied to the lower cycle is

$$Q_L = Q_{HL} - \lambda_D$$

$$= \frac{1 - \eta_H(1 + \lambda_D)}{(\eta_H + \eta_L - \eta_H \eta_L)}. \tag{3.39}$$

Q_L becomes zero when $\eta_H = \dfrac{1}{(1 + \lambda_D)}$ i.e. when $\lambda_D = \left(\dfrac{1 - \eta_H}{\eta_H}\right)$.

No bottoming cycle then exists, because the topping cycle is exactly matched to the heat and electricity loads and becomes a "matched" back-pressure set. The energy utilisation factor is then unity.

(iii) If $\lambda > \left(\dfrac{1 - \eta_H}{\eta_H}\right)$, and the electrical demand continues to be met, then the heat and electrical work (Q_L and W_L) both become negative. Essentially the lower cycle becomes a heat pump, with the heat load being met by the sum of rejected heat Q_{HL} from the upper cycle, and heat pumped from the lowest (ambient) temperature T_0 to the required useful temperature T_U. This type of operation is discussed further below in section 3.4.4.

In the example of the back-pressure turbine (of efficiency η_{CG}) its operation was considered in parallel with a conventional power plant, the two together being analysed as a combined plant. Here a conventional plant is essentially "hidden" within the plant indicated in Fig. 3.4. This may be illustrated by developing the analysis further.

If the combined power plant efficiency ($\eta_H + \eta_L - \eta_H\eta_L$) is the same as that of a conventional plant (η_C) (i.e. if steam is being tapped off from a conventional plant), then the lower efficiency is

$$\eta_L = \frac{\eta_C - \eta_H}{(1 - \eta_H)}. \tag{3.40}$$

Equations (3.37) and (3.38) may then be written as

$$(\text{EUF})' = \frac{(1 + \lambda_D)\eta_C}{1 + \left(\dfrac{\eta_C - \eta_H}{1 - \eta_H}\right)\lambda_D} = \frac{(1 - \eta_H)(1 + \lambda_D)\eta_C}{(1 - \eta_H) + \lambda_D(\eta_C - \eta_H)} \tag{3.41}$$

$$(\text{FESR})' = 1 - \frac{[(1 - \eta_H) + (\eta_C - \eta_H)\lambda_D]\eta_B}{(\eta_B + \eta_C\lambda_D)(1 - \eta_H)}. \tag{3.42}$$

These expressions are identical with equations (3.17) and (3.18) if the efficiency of the upper plant here (η_H) is the same as that of the back-pressure turbine (η_{CG}) in section 3.4. In other words, the plant of Fig. 3.4 may be considered as a conventional combined power plant plus a back-pressure plant as indicated in Fig. 3.5.

We assumed in this analysis that the steam extraction plant supplied heat and power to match the demand requirements (λ_D). In fact, such an extraction plant can be designed for a range of λ_D, by designing for different amounts of steam extraction (essentially changing the hidden balance between the conventional and back-pressure plants). Further a plant designed to meet a particular value of λ_D can then to some extent operate "off-design" if λ_D changes with time; this off-design operation is a matter for further discussion later.

The simple type of analysis given for this plant may be modified by allowing for the heat loss ($\phi\lambda_D$) in supplying the heat load. Equations (3.33) and (3.34) become then

$$Q_{HL} = Q_L + (1 + \phi)\lambda_D, \tag{3.43}$$

and

$$W = 1 = \eta_H F' + \eta_L Q_L$$

$$= F'[\eta_H + \eta_L(1 - \eta_H)] - \eta_L(1 + \phi)\lambda_D. \tag{3.44}$$

The performance parameters are then

$$\eta' = \frac{\eta_H + \eta_L - \eta_H\eta_L}{1 + \eta_L(1 + \phi)\lambda_D}, \tag{3.45}$$

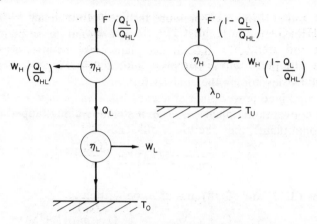

FIG. 3.5. Combined plant of Fig. 3.4, broken down
as conventional combined plant plus back pressure
plant.

$$(EUF)' = \frac{(1 + \lambda_D)(\eta_H + \eta_L - \eta_H\eta_L)}{1 + \eta_L(1 + \phi)\lambda_D}, \qquad (3.46)$$

$$(FESR)' = 1 - \frac{[1 + \eta_L(1 + \phi)\lambda_D](\eta_C\eta_B)}{(\eta_H + \eta_L - \eta_H\eta_L)(\eta_B + \eta_C\lambda_D)}. \qquad (3.47)$$

We return to more detailed discussion of the pass-out or extraction steam turbine in section 3.8.2, where we relate the amount of bled steam to the useful heat load. We shall also discuss the combination of feed heating and district heating, both from bled steam.

3.4.3 Gas Turbine with Waste Heat Recuperator or Boiler (Heat Recovery Steam Generator (HRSG) Fired or Unfired)

The back-pressure steam plant of section 3.4.1 rejected all its heat in a useful form. A gas turbine (open circuit or closed circuit) with a WHR (waste heat recuperator) (example H of Fig. 2.1) does not reject all its heat usefully. In an open circuit plant exhaust gases from the recuperator leave at a temperature above that of the environment (condensation must be avoided in the exhaust stack); further, additional fuel may be burnt in a waste heat boiler for the turbine exhaust gases still contain surplus oxygen. In a closed circuit plant, before the gas re-enters the compressor it usually passes through an "internal" heat exchanger, a waste heat recuperator and a "pre-cooler" (see Bammert[2] who describes such closed-circuit gas turbine plant, and Campbell et al.[3]).

Figure 3.6(a) shows a block diagram of a "matched" gas turbine with a WHR (HRSG, unfired); this CHP plant (subscript CG) supplies power $W_{CG} = 1$ and useful heat $(Q_U)_{CG} = \lambda_{CG} = \lambda_D$ and rejects non-useful heat $(Q_{NU})_{CG}$. The performance parameters are those given in section 3.3 for the "matched" CHP plant,

$$\eta_{CG} = \frac{1}{1 + \lambda_{CG} + (Q_{NU})_{CG}}, \tag{3.6}$$

and
$$\lambda_{CG} = \left(\frac{1}{\eta_{CG}} - 1\right) \Big/ \left[1 + \left(\frac{Q_{NU}}{\lambda}\right)_{CG}\right],$$

$$(EUF)_{CG} = \frac{1 + \lambda_{CG}}{1 + \lambda_{CG} + (Q_{NU})_{CG}}, \tag{3.7}$$

$$(FESR)_{CG} = 1 - \frac{\eta_C \eta_B}{\eta_{CG}(\eta_B + \lambda_{CG}\eta_C)}. \tag{3.9}$$

The quantity $(Q_{NU})_{CG}$ is usually fixed by the allowable stack temperature T_S, i.e. $(Q_{NU})_{CG} = m_G[(h_G)_s - (h_G)_0]$ where m_G is the mass of exhaust gas entering the stack with specific enthalpy $(h_G)_s$, per unit of work produced.

If the CHP plant is not matched, meeting the heat load $((Q_U)_D = (Q_U)_{CG})$ but not the power load $(W_{CG} < W_D)$ then extra power from the grid is required (W_C). Following a procedure similar to that given in section 3.4.1, it is straightforward to show that the performance parameters for the total plant are then

$$\eta' = \frac{(W_C + W_{CG})}{F'} = \frac{(1 - \eta_{CG})\eta_C}{(1 - \eta_{CG}) + \left[1 + \left(\frac{(Q_{NU})_{CG}}{\lambda_D}\right)\right](\eta_C - \eta_{CG})\lambda_D}$$

$$\tag{3.48}$$

$$EUF' = \frac{(1 + \lambda_D)}{F'} = \frac{(1 + \lambda_D)(1 - \eta_{CG})\eta_C}{(1 - \eta_{CG}) + \left[1 + \left(\frac{(Q_{NU})_{CG}}{\lambda_D}\right)\right](\eta_C - \eta_{CG})\lambda_D}$$

$$\tag{3.49}$$

$$FESR' = 1 - \left(\frac{F'}{F_{REF}}\right)$$

$$= 1 - \eta_B \left[\frac{(1 - \eta_{CG}) + \left[1 + \left(\frac{(Q_{NU})_{CG}}{\lambda_D}\right)\right](\eta_C - \eta_{CG})\lambda_D}{(1 - \eta_{CG})(\eta_B + \eta_C\lambda_D)}\right]$$

$$\tag{3.50}$$

(a) Matched plant with waste heat recuperator (heat recovery steam generator, unfired)

(b) Unmatched plant with $(Q_U)_{CG} = (Q_U)_D$, but $W_{CG} < W_D$

(c) Matched plant with waste heat boiler (heat recovery steam generator fired)

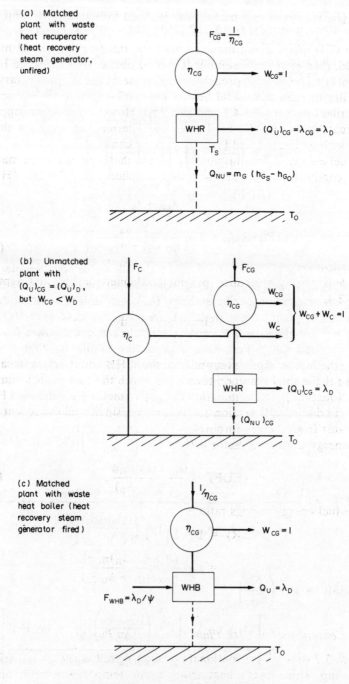

FIG. 3.6. Gas turbine CHP plant.

where $(Q_{NU})_{CG}$ is the non-useful heat rejected by the cogeneration plant alone.

If the CHP plant is not matched, meeting the power load but not the heat load, then increased useful heat may be obtained by firing the HRSG (Fig. 3.6(c)). This is an approach similar to use of the supplementary heat only boiler to raise the useful heat of the back-pressure turbine scheme (as described in section 3.4.1 and Fig. 3.3). However, here we suppose a waste heat boiler (HRSG fired) uses the internal exhaust from the gas turbine with additional fuel (usually gas) for combustion.

The fuel energy supplied is now the sum of that supplied to the main gas turbine combustion chamber and that supplied to the WHB (HRSG, fired),

$$F' = F_{CG} + F_{WHB}$$

$$= \frac{1}{\eta_{CG}} + \frac{\lambda_D}{\psi} \tag{3.51}$$

where ψ is the performance parameter defined earlier in Chapter 2, section 2.4,

$$\psi = \frac{\text{Heat transferred}}{m_f(CV)_0},$$

m_f being the mass of fuel gas supplied to the WHB and $(CV)_0$ in its calorific value for the ambient temperature T_0, at which the fuel gas is assumed to enter. (Note that the determination of the parameter ψ for the fired HRSG (WHB), as discussed in section 2.4, is complex, since it takes account of the hot exhaust from the gas turbine).

The energy utilisation factor is then

$$(EUF)' = \frac{(\lambda_D + 1)\eta_{CG}\psi}{(\psi + \eta_{CG}\lambda_D)}, \tag{3.52}$$

and the fuel energy savings ratio is

$$(FESR)' = 1 - (F'/F_{REF})$$

$$= 1 - \frac{(\psi + \eta_{CG}\lambda_D)\eta_C\eta_B}{\eta_{CG}\psi(\eta_B + \lambda_D\eta_C)}. \tag{3.53}$$

3.4.4 Conventional or CHP Plant Driving Heat Pump

Figure 3.7 shows a conventional plant, of overall efficiency η_C, driving a heat pump which takes heat from a low temperature reservoir (the atmosphere at T_0) and, receiving work W_{HP}, pumps heat $(Q_U)_{HP} = \lambda_D$ to

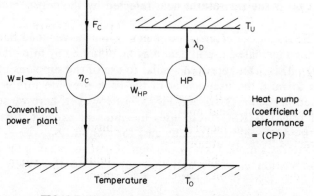

FIG. 3.7. Heat pump driven by conventional power plant.

a higher temperature to meet the heat load. This scheme operates for large values of λ_D as we discovered in section 3.4.2.

The heat pump's performance is defined by the coefficient of performance (CP),

$$(CP) = \frac{\lambda_D}{W_{HP}}. \tag{3.54}$$

The equations for overall performance are now

$$W_C = W_{HP} + 1 = \eta_C F_C, \tag{3.55}$$

and

$$W_{HP} = \lambda_D/(CP). \tag{3.56}$$

Hence

$$\eta_C F_C = [\lambda_D/(CP)] + 1, \tag{3.57}$$

and

$$F_C = \frac{(CP) + \lambda_D}{\eta_C(CP)}. \tag{3.58}$$

The overall efficiency of the total scheme is

$$\eta' = \frac{1}{F_C} = \frac{\eta_C(CP)}{(CP) + \lambda_D}, \tag{3.59}$$

but has no real significance.

The energy utilisation factor is

$$(EUF)' = \frac{(1 + \lambda_D)}{F_C} = \frac{\eta_C(1 + \lambda_D)(CP)}{(CP) + \lambda_D}, \tag{3.60}$$

and the fuel energy savings ratio is

$$(\text{FESR})' = 1 - (F_C/F_{\text{REF}}) = 1 - \frac{\eta_B[1 + (\lambda_D/(CP))]}{(\eta_B + \eta_C\lambda_D)}. \tag{3.61}$$

The following points may be noted.

(i) As $\lambda_D \to \infty$ so $\eta' \to 0$, $(\text{EUF})' \to \eta_C(CP)$.
(This is essentially a "heat only" scheme.)

(ii) The energy utilisation factor can exceed unity in this case since the heat extracted from the atmosphere is "free". This illustrates that this criterion of performance has a restricted usefulness, as a value of unity does not always represent "perfect" performance.

Losses in the conversion of the pumped heat into the heat load may be allowed for by replacing λ_D by $\lambda_D(1 + \phi)$, so that the heat losses are $\phi\lambda_D$. The efficiency, EUF and FESR are then

$$\eta' = \eta_C/\{1 + [\lambda_D(1 + \phi)/(CP)]\}, \tag{3.62}$$

$$(\text{EUF})' = \frac{\eta_C(1 + \lambda_D)}{1 + [\lambda_D(1 + \phi)/(CP)]}, \tag{3.63}$$

$$(\text{FESR})' = 1 - \frac{\eta_B\{1 + [\lambda_D(1 + \phi)/(CP)]\}}{(\eta_B + \eta_C\lambda_D)}$$

$$= \frac{\lambda_D\{\eta_C - [\eta_B(1 + \phi)/(CP)]\}}{\eta_B + \eta_C\lambda_D}. \tag{3.64}$$

The use of a heat pump in a "heat only" scheme has already been mentioned (scheme K of Fig. 2.1, where reference was made to the proposal by Kolbusz[4], in which the heat load was met partially by heat rejection from a back-pressure turbine and partially from a heat pump). Before considering that option we discuss further the case of the heat pump solely meeting the heat load and driven by a conventional power plant (i.e. through the electricity grid).

In this case (Fig. 3.8(a)) there is no work output, so that $W_C = W_{HP}$. The heat supplied by the heat pump is $(Q_U)_{HP}$; if the heat load is $(Q_U)_D$ and there are heat losses $(\phi(Q_U)_D)$ on the heat delivery side, then $(Q_U)_{HP} = (Q_U)_D(1 + \phi)$. The fuel input is then $F_C = (Q_U)_D(1 + \phi)/\eta_C(CP)$ and the performance parameters are

$$(\text{EUF})' = \frac{(Q_U)_D}{F_C} = \frac{\eta_C(CP)}{(1 + \phi)}, \tag{3.65}$$

$$(\text{FESR})' = 1 - (F_C/F_{\text{REF}}) = 1 - \frac{(1 + \phi)\eta_B}{\eta_C(CP)}, \tag{3.66}$$

(a)

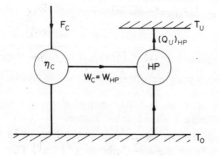

(b)

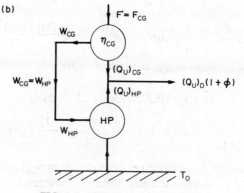

FIG. 3.8. "Heat only" heat pump schemes.

since $F_{REF} = (Q_U)_D/\eta_B$.

(These expressions correspond to the case of $\lambda_D \to \infty$ in the scheme of Fig. 3.7.) There is clearly no fuel saving associated with such a "heat only" scheme unless

$$(CP) > \eta_B(1 + \phi)/\eta_C,$$

which is a difficult criterion to meet.

The proposal by Kolbusz, with the heat load shared between the heat pump and the back-pressure turbine driving it (Fig. 3.8(b)), may ease this requirement of high performance from the heat pump. If the efficiency of the turbine is η_{CG} and the fuel energy supplied to it is F_{CG} then the relevant equations are

$$\eta_{CG}F_{CG} = W_{CG}, \tag{3.67}$$

$$(Q_U)_{CG} = F_{CG} - W_{CG} = F_{CG}(1 - \eta_{CG}), \tag{3.68}$$

$$(Q_U)_{HP} = (CP)W_{HP} = (CP)W_{CG}$$
$$= F_{CG}\eta_{CG}(CP), \tag{3.69}$$

$$(Q_U)_D(1 + \phi) = (Q_U)_{CG} + (Q_U)_{HP}$$
$$= F_{CG}(1 - \eta_{CG} + \eta_{CG}(CP)). \tag{3.70}$$

It follows that

$$F_{CG} = \frac{(Q_U)_D(1 + \phi)}{1 + \eta_{CG}[(CP) - 1]}, \tag{3.71}$$

$$(EUF)' = \frac{(Q_U)_D}{F_{CG}} = \frac{1 + \eta_{CG}[(CP) - 1]}{(1 + \phi)}, \tag{3.72}$$

$$(FESR)' = 1 - (F_{CG}/F_{REF}) = 1 - \frac{\eta_B(1 + \phi)}{1 + \eta_{CG}[(CP) - 1]}. \tag{3.73}$$

The requirement for fuel savings is now

$$1 + \eta_{CG}[(CP) - 1] > (1 + \phi)\eta_B,$$

or
$$(CP) > 1 + \frac{[(1 + \phi)\eta_B - 1]}{\eta_{CG}}. \tag{3.74}$$

The requirement on heat pump performance is now much less stringent—roughly that CP must exceed unity, virtually independent of η_{CG}, for fuel savings to be achieved.

An even more complex scheme has been studied by Casci and Gaia[5], who propose a cogeneration plant comprising a gas turbine with bottoming condensing cycle (using organic fluid) *and* a heat pump. The "district" heating (at proposed temperatures which may be too low to be practical) is achieved through heat rejection from the bottoming organic fluid cycle and heat supplied from the heat pump, which pumps heat out of the gas turbine exhaust, requiring it to leave at a very low (ambient) temperature of 20°C.

3.5 CHP Plants Operating on Carnot Cycles

It is instructive to assume that the components in each of the CHP plant described operate on Carnot cycles, since this enables the performance of three of the four plants studied so far to be described by a single analysis (see Horlock[6]). If the fuel energy supplies (F) of Fig. 3.2 are replaced by heat supplies (Q) from reservoirs at temperature T_B, useful heat is supplied at T_U and heat rejected at T_0, then the various Carnot efficiencies are simply expressed.

Thus, for the example of Fig. 3.2

$$\eta_C = \left(\frac{T_B - T_0}{T_B}\right), \text{ for the conventional plant,} \tag{3.75}$$

and

$$\eta_{CG} = \left(\frac{T_B - T_U}{T_B}\right), \text{ for the cogeneration plant.} \tag{3.76}$$

It follows from equations (3.15), (3.16), (3.17) and (3.18) that

$$Q' = \frac{1}{\eta_C} + \frac{\lambda_D}{(1 - \eta_{CG})}\left(1 - \frac{\eta_{CG}}{\eta_C}\right) \tag{3.77a}$$

$$= \frac{T_B(1 + \lambda_D - \lambda_D T_0/T_U)}{T_B - T_0}, \tag{3.77b}$$

$$\eta' = \frac{(1 - \eta_{CG})\eta_C}{(1 - \eta_{CG}) + \lambda_D(\eta_C - \eta_{CG})} = \frac{T_B - T_0}{T_B(1 + \lambda_D - \lambda_D T_0/T_U)}, \tag{3.78}$$

$$(EUF)' = \frac{(T_B - T_0)(1 + \lambda_D)}{T_B(1 + \lambda_D - \lambda_D T_0/T_U)}, \tag{3.79}$$

$$(FESR)' = [(T_B/T_U) - 1]\bigg/\left[\left(\frac{T_B}{T_0}\right)\left(\frac{1 + \lambda_D}{\lambda_D}\right) - 1\right], \tag{3.80}$$

assuming that an ideal boiler is available to supply heat at T_U in the reference case. For the example of Fig. 3.4 (the pass-out turbine),

$$\eta_H = \left(\frac{T_B - T_U}{T_B}\right), \eta_L = \left(\frac{T_U - T_0}{T_U}\right), \quad \text{so that equation (3.35)}$$

$$F' = \frac{(1 + \eta_L\lambda_D)}{(\eta_H + \eta_L - \eta_H\eta_L)}, \tag{3.35}$$

becomes

$$Q' = \frac{T_B}{(T_B - T_0)}\left(1 + \lambda_D - \frac{\lambda_D T_0}{T_U}\right), \tag{3.81}$$

which is the same as equation (3.77(b)). The overall efficiency, $(EUF)'$ and $(FESR)'$ are obviously also the same, given by equations (3.78), (3.79) and (3.80).

For the example of Fig. 3.8(a) a conventional plant driving a heat pump,

$$(CP) = \frac{T_U}{T_U - T_0}, \tag{3.82}$$

so that

$$Q' = \frac{[1 + \lambda_D/(CP)]}{\eta_C} = \frac{T_B}{(T_B - T_0)}\left[1 + \lambda_D - \frac{\lambda_D T_0}{T_U}\right], \tag{3.83}$$

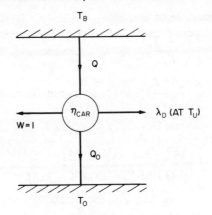

FIG. 3.9. Carnot engine CHP
scheme.

which again is the same as equation (3.77(b)). The overall efficiency, (EUF)′
and (FESR)′ are again given by equations (3.78), (3.79) and (3.80).

That the expressions for the performance criteria for each example are
identical is not, in retrospect, surprising. All three systems are now fully
reversible and essentially identical.

A simple analysis (Haywood[7]) enables the equation (3.77(b)) to be
derived immediately. Figure 3.9 shows a fully reversible system supplying
unit electrical work and heat λ_D at temperature T_U. Heat Q_0 is rejected at
temperature T_0.

From the first law of thermodynamics, $Q - \lambda_D - Q_0 = 1$. From the
second law (since the system is fully reversible),

$$\frac{Q}{T_B} - \frac{\lambda_D}{T_U} - \frac{Q_0}{T_0} = 0. \tag{3.84}$$

From these two equations Q_0 may be eliminated to give equation (3.77(b)),

$$Q = \frac{T_B}{(T_B - T_0)}\left[1 + \lambda_D - \frac{\lambda_D T_0}{T_U}\right]. \tag{3.85}$$

3.6   Numerical Examples

3.6.1   CHP Plants Operating on Carnot Cycles

The advantage of the unified (reversible) analysis of the last section
enables the performance criteria to be simply determined as functions of
λ_D, if T_B, T_U and T_0 are specified, for the three forms of operation
described in section 3.4.

If the temperatures are taken as $T_B = 1200 \text{ K}$, $T_U = 400 \text{ K}$, $T_0 = 300 \text{ K}$ then the following performance parameters are obtained:

$$\eta_C = 3/4, \tag{3.86}$$

$$\eta_{CG} = 2/3, \tag{3.87}$$

$$Q' = (\lambda_D + 4)/3, \tag{3.88}$$

$$\eta' = 3/(\lambda_D + 4), \tag{3.89}$$

$$(EUF)' = 3(\lambda_D + 1)/(\lambda_D + 4), \tag{3.90}$$

$$(FESR)' = 2\lambda_D/(4 + 3\lambda_D), \tag{3.91}$$

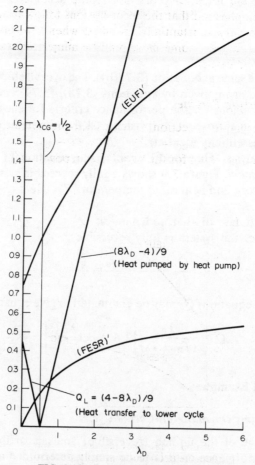

FIG. 3.10. Carnot engine CHP schemes $(\lambda_{CG} = \tfrac{1}{2})$.

where the reference plant is based on a Carnot cycle. Figure 3.10 shows
(EUF)′ and (FESR)′ plotted against the heat/electricity power ratio (λ_D),
together with the heat transferred to the lower cycle in the pass-out
turbine of Fig. 3.4,

$$Q_L = (4 - 8\lambda_D)/9. \tag{3.92}$$

This expression shows that the critical value of λ_D is $\lambda_{CG} = \frac{1}{2}$ when $Q_L = 0$.
For $\lambda_D < \frac{1}{2}$, either the pass-out turbine (with a grid turbine supplying extra
power) or the back-pressure turbine (thermodynamically identical, as
shown in Fig. 3.5) operate. For $\lambda_D > \frac{1}{2}$, the heat pump plant operates,
pumping heat $(8\lambda_D - 4)/9$.

It would appear from this ideal analysis that for $\lambda_D < \frac{1}{2}$ the turbines are
the best choice and for $\lambda_D > \frac{1}{2}$ the heat pump scheme is the optimum.
But it must be emphasised that these conclusions follow from a reversible
analysis and are very substantially modified when real plants are considered. We now consider some more realistic numerical examples for the
four plants studied earlier.

3.6.2 *Back-Pressure Turbine*

For the first example of section 3.4 (the back-pressure turbine) we take
Timmermans' values of $\eta_{CG} = 0.25$ ($\lambda_{CG} = (1 - \eta_{CG})/\eta_{CG} = 3$) for the
back-pressure set, $\eta_C = 0.4$ for the "grid" steam power plant, and $\eta_B = 0.9$

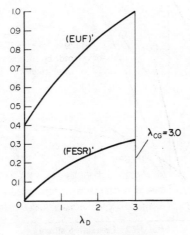

FIG. 3.11. Back pressure
turbine (CG) plus conventional
power plant (C). $\eta_{CG} = 0.25$,
$\eta_C = 0.4$, $(\eta_B)_C = 0.9$.

for the original boiler plant. Equations (3.17) and (3.18) then lead to the following expressions:

$$(EUF)' = \frac{0.4(1 + \lambda_D)}{1 + 0.2\lambda_D}, \tag{3.93}$$

$$(FESR)' = 1 - \frac{0.9(0.75 + 0.15\lambda_D)}{0.75(0.9 + 0.4\lambda_D)} = \frac{0.22\lambda_D}{0.9 + 0.4\lambda_D}, \tag{3.94}$$

which are plotted in Fig. 3.11.

The value of $(EUF)'$ rises from 0.4 at $\lambda_D = 0$ to 1.0 at the critical value of $\lambda_{CG} = 3.0$ (when the back-pressure turbine operates alone). $(FESR)'$ also rises steadily with λ_D up to a value of 0.31 at $\lambda_{CG} = 3.0$.

3.6.3   Pass-Out or Extraction Turbine

For the second example of section 3.4 (the extraction turbine) if we take $\eta_H = 0.333$ for the topping plant), and $\eta_L = 0.1$ for the bottoming plant then the overall combined power plant efficiency $(\eta_H + \eta_L - \eta_H\eta_L) = 0.4$. It is also assumed that the efficiency of the original conventional plant is 0.4, and with $\eta_B = 0.9$, the following expressions may be derived from equations (3.37) and (3.38):

$$(EUF)' = \frac{0.4(1 + \lambda_D)}{1 + 0.1\lambda_D}, \tag{3.95}$$

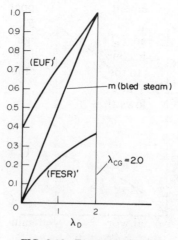

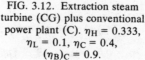

FIG. 3.12. Extraction steam
turbine (CG) plus conventional
power plant (C). $\eta_H = 0.333$,
$\eta_L = 0.1$, $\eta_C = 0.4$,
$(\eta_B)_C = 0.9$.

$$(FESR)' = 1 - \frac{0.9(1 + 0.1\lambda_D)}{0.9 + 0.4\lambda_D} = \frac{0.31\lambda_D}{0.9 + 0.4\lambda_D}, \qquad (3.96)$$

which are plotted in Fig. 3.12.

$(EUF)'$ rises from 0.4 to 1.0 at the critical value of $\lambda_{CG} = \left(\dfrac{1 - \eta_H}{\eta_H}\right) = 2.0$.

$(FESR)'$ rises steadily with λ_D, up to 0.365 at $\lambda_{CG} = 2.0$.

If we had taken the topping plant efficiency as 0.25 (the same as the back-pressure plant of section 3.8.2) and that of the bottoming plant as 0.2 (a high value) then the efficiency of the combined plant would still be 0.4. However, equations (3.95) and (3.96) would have become identical to (3.93) and (3.94) and the performance would be as illustrated by Fig. 3.11. Later in section 3.8 we interpret Fig. 3.12 in terms of the amount of bled steam required.

3.6.4   Gas Turbine with Waste Heat Recuperator or Boiler (Heat Recovery Steam Generator, Fired or Unfired)

For the third example, of section 3.4.3 (gas turbine with waste heat recuperator) we take a value of $\eta_{CG} = 0.25$ for a gas turbine (somewhat less than Timmermans' value), with $\eta_C = 0.4$ as the original basic plant efficiency and $\eta_B = 0.9$. If $(Q_{NU})_{CG}/(\lambda)_{CG} = 1/3$, then from equation (3.6) the plant is matched with

$$\lambda_D = \lambda_{CG} = \left(\frac{1}{\eta_{CG}} - 1\right) \bigg/ [1 + (Q_{NU})_{CG}] = \frac{3}{(4/3)} = 2.25.$$

The values of $(EUF)_{CG}$ and $(FESR)_{CG}$ (from equations (3.7) and (3.8)) are then

$$(EUF)_{CG} = \frac{3.25}{1 + 2.25(4/3)} = \frac{3.25}{4} = 0.812,$$

$$(FESR)_{CG} = 1 - \frac{0.4 \cdot 0.9}{0.25(0.9 + 2.25 \cdot 0.4)} = 0.20.$$

For the unmatched plant (with recuperator) meeting the heat load, but with extra electricity bought from the grid,

$$(EUF)' = \frac{(1 + \lambda_D)(0.75 \cdot 0.4)}{0.75 + \lambda_D(4/3)0.15} = \frac{0.3 + 0.3\lambda_D}{0.75 + 0.2\lambda_D}, \qquad (3.97)$$

$$(FESR)' = 1 - \frac{0.9(0.75 + (4\lambda_D/3)0.15)}{0.75(0.9 + 0.4\lambda_D)} = \frac{0.12\lambda_D}{0.675 + 0.3\lambda_D}.$$

$$(3.98)$$

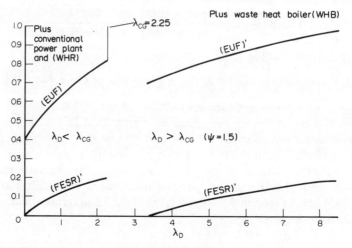

FIG. 3.13. Gas turbine.

These parameters are shown plotted against λ_D on the left-hand side of Fig. 3.13.

For the unmatched plant, with power demand met and a waste heat boiler (HRSG fired) to meet the extra heat load, if $\psi = 1.5$ then from equations (3.52) and (3.53) the performance parameters become

$$(\text{EUF})' = \frac{(\lambda_D + 1)0.25 \cdot 1.5}{1.5 + 0.25\lambda_D} = \frac{1.5(1 + \lambda_D)}{6 + \lambda_D}, \qquad (3.99)$$

$$(\text{FESR})' = 1 - \frac{(1.5 + 0.25\lambda_D)0.4 \cdot 0.9}{1.5 \cdot 0.25(0.9 + 0.4\lambda_D)} = \frac{0.24\lambda_D - 0.81}{1.35 + 0.6\lambda_D} \qquad (3.100)$$

(but note that the practical plant of section 5.4 has a somewhat lower value of ψ, equal to 1.34).

These parameters are shown plotted against λ_D on the right-hand side of Fig. 3.13. Fuel savings appear only for $\lambda_D > 3.37$.

3.6.5 Conventional Power Plant Driving a Heat Pump

For the fourth example, of section 3.4.4 (conventional power plant driving a heat pump) we take $\eta_C = 0.4$, the coefficient of performance (CP) as 2.2, the loss factor as 0.1, and $\eta_B = 0.9$. From equations (3.63) and (3.64) the expressions for $(\text{EUF})'$ and $(\text{FESR})'$ are

$$(\text{EUF})' = \frac{0.4(1 + \lambda_D)}{1 + 0.5\lambda_D}, \qquad (3.101)$$

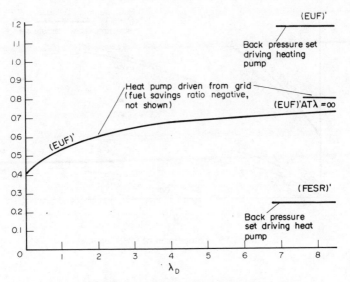

FIG. 3.14. Heat pump schemes.

$$(\text{FESR})' = -\frac{0.05\lambda_D}{0.9 + 0.4\lambda_D}. \qquad (3.102)$$

Although the energy utilisation factor increases from 0.4 at $\lambda_D = 0$ to 0.8 at very large λ_D (see Fig. 3.14) the fuel energy savings ratio is always negative.

Even for the heat only scheme (corresponding to $\lambda_D = \infty$) in which the heat pump is simply driven by a conventional plant and no external power is supplied, with the same assumptions

$$(\text{EUF})' = 0.4/0.5 = 0.8,$$

$$(\text{FESR})' = \frac{0.4 - 0.45}{0.4} = -0.125,$$

and there are no fuel savings. (CP) must exceed $\eta_B(1 + \phi)/\eta_C$ for fuel savings to occur; thus if $\eta_B = 0.9$, $\phi = 0.1$ and $\eta_C = 0.4$, the coefficient of performance has to exceed 2.5.

However, for the heat only scheme proposed by Kolbusz and described in section 3.4.4 (back-pressure turbine of efficiency η_{CG}, rejecting useful heat and driving the heat pump) with the same assumptions for (CP), ϕ, η_B and with $\eta_{CG} = 0.25$, from equations (3.72) and (3.73)

$$(\text{EUF})' = \frac{1 + 0.25 \cdot 1.2}{1.1} = 1.17,$$

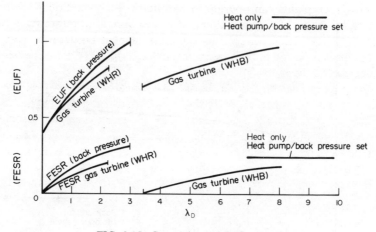

FIG. 3.15. Comparison of CHP schemes.

and
$$(FESR)' = 1 - \frac{1.1 \cdot 0.9}{(1 + 0.25 \cdot 1.2)} = 0.239,$$

also for $\lambda_D = \infty$.

There are now fuel savings since the criterion

$$(CP) > 1 + \left(\frac{(1 + \phi)\eta_B - 1}{\eta_{CG}} \right) = 0.975$$

has been met.

3.6.6 *Comparison of Performance Criteria*

The numerical calculations of Figs 3.11, 3.13 and 3.14 are put together in Fig. 3.15. They illustrate the advantages of back-pressure (or extraction) turbines at low values of the heat to power ratio λ_D and the advantages of the gas turbine (with a supplementary boiler) and a heat pump/back-pressure set at high λ_D.

3.7 Calculations of Typical Heat to Power Ratios (λ_{CG}) and Performance Parameters for "Self-Contained" CHP Plant

The analysis of section 3.4 and the calculations of section 3.6 have used assumed values of CHP plant efficiencies (and other parameters) to illustrate how such plants can operate in parallel with conventional power or boiler plants, to meet specified demand values of heat to power ratio (λ_D). If $\lambda_D = \lambda_{CG}$, the ratio produced by the CHP plant itself, then no conventional plant is required to operate in parallel, i.e. the CHP plant is perfectly matched to the demands.

In this section, calculated values of λ_{CG} associated with particular CHP plants are given, together with values of performance criteria. Such calculations, illustrating how each type of plant has an associated (approximate) value of λ_{CG} have been made by Porter and Mastanaiah[1] using their definition of incremental heat rate (equation (2.9)),

$$(\text{IHR})_{CG} = \frac{F_{CG}}{W_{CG}} - \frac{Q_U}{\eta_B W_{CG}} = \frac{1}{\eta_{CG}} - \frac{\lambda_{CG}}{(\eta_B)_{CG}}. \qquad (2.9)$$

They determine from simple thermodynamic analyses how $(\text{IHR})_{CG}$ and the ratio $\lambda_{CG} = (Q_U/W)_{CG}$ vary with some thermodynamic design parameters, for three types of plant: a back-pressure steam turbine; and a gas turbine and a Diesel engine, each with a waste heat exchanger supplying the heat load. (See Figs 3.16, 3.17, 3.18.)

In the first example (a "matched" back-pressure turbine) there is no heat rejection other than the useful heat supplied to the process plant. Thus with $(Q_{NU})_{CG} = 0$ in equation (2.7),

$$F_{CG} = \frac{(Q_U + W)_{CG}}{\eta_B}. \qquad (3.103)$$

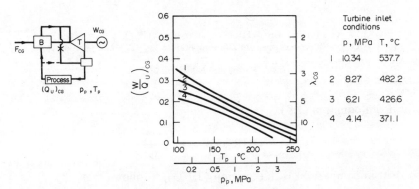

FIG. 3.16. Back pressure turbine—variation of $(W/Q_U)_{CG}$ ($= 1/\lambda_{CG}$) with process pressure and temperature (p_p, T_p).

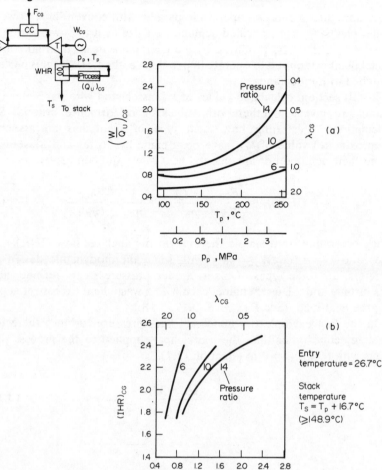

FIG. 3.17. Gas turbine with WHR—variation of (a) $(W/Q_U)_{CG}$ (= $1/\lambda_{CG}$) and (b) $(IHR)_{CG}$ with process pressure and temperature (p_p, T_p), and pressure ratio. Maximum temperature = 927°C, $\eta_{cc} = 0.95$, $\eta_T = 0.9$, $\eta_C = 0.9$ (after Porter and Mastanaiah[1]).

But, from equation (2.9),

$$F_{CG} = (IHR)_{CG} W_{CG} + \frac{(Q_U)_{CG}}{\eta_B},$$

so that, in comparison with equation (3.103), it follows that

$$(IHR)_{CG} = 1/(\eta_B)_{CG}, \qquad (3.104)$$

which is little greater than unity.

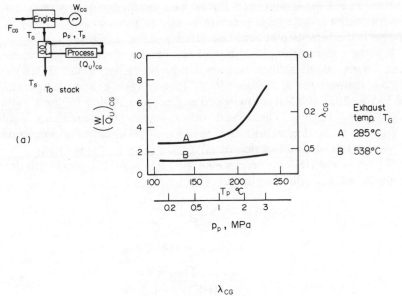

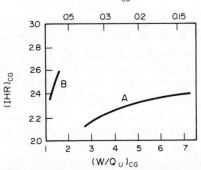

FIG. 3.18. Diesel engine with WHR—variation of $(W/Q_U)_{CG}$ $(= 1/\lambda_{CG})$ and $(IHR)_{CG}$ with process pressure and temperature (p_p, T_p). Stack temperature $T_S \geqslant 177°C$ (after Porter and Mastanaiah[1]).

Porter and Mastanaiah also allow for a generator efficiency η_g, which leads to $(IHR)_{CG} = (1/(\eta_B)_{CG}\eta_g)$. With $\eta_g = 0.96$ and $(\eta_B)_{CG} = 0.85$ they take $(IHR)_{CG} = 1.23$ as a representative figure for a back-pressure steam turbine plant. For a plant providing saturated process steam at pressure p_p, temperature T_p, they derive a plot of $(W/Q_U)_{CG}$ against T_p, for a turbine efficiency of 0.75, $(IHR)_{CG} = 1.23$ and various turbine entry conditions (Fig. 3.16).

In their second example of a gas turbine with a waste heat recuperator (WHR) supplying process steam at p_p, T_p, plots of $(W/Q_U)_{CG}$ $(= 1/\lambda_{CG})$

against T_p are again produced for various thermodynamic parameters of the gas turbine cycle (e.g. pressure ratio), as given in Fig. 3.17(a); also shown is the incremental heat rate (IHR)$_{CG}$ (Fig. 3.17(b)).

For their third example of the Diesel engine, Porter and Mastanaiah take a more empirical approach, assuming a waste heat boiler (WHB) with a stack temperature of 350°F (176.6°C); and engine exhaust temperatures of 545°F (285.1°C) (for a low-speed engine) and 1000°F (537.8°C) (for a high speed engine. These and other empirical assumptions enable $(W/Q_U)_{CG}$ to be determined as a function of process steam temperature T_p, and (IHR)$_{CG}$ as a function of $(W/Q_U)_{CG}$ ($= 1/\lambda_{CG}$) (Fig. 3.18).

With λ_{CG} and (IHR)$_{CG}$ determined for each plant it is possible to determine the energy utilisation factor

$$
\begin{aligned}
(EUF)_{CG} &= \frac{(Q_U + W)_{CG}}{F_{CG}} \\
&= (\lambda_{CG} + 1)W_{CG}/F_{CG} \\
&= \frac{(\lambda_{CG} + 1)}{(IHR)_{CG} + (\lambda_{CG}/\eta_B)}.
\end{aligned}
\tag{3.105}
$$

Figure 3.19 shows the results of Porter and Mastanaiah for the three cycles they studied, in the form of (EUF)$_{CG}$ plotted against λ_{CG} for specified values of (IHR)$_{CG}$ and $(\eta_B)_{CG}$ (Fig. 3.19a). The areas of operation of the back-pressure turbine, gas turbine and Diesel engine are all shown. The back-pressure steam turbine shows the highest (EUF)$_{CG}$.

Gas turbines show good energy utilisation at λ_{CG} near unity, and Diesel engines the lowest (EUF)$_{CG}$, at low λ_{CG}. Porter and Mastanaiah also give values of the corresponding fuel energy savings ratio, (FESR)$_{CG}$ (Fig. 3.19(b)).

While these calculations give a useful guide to how the choice of a "self-contained" CHP plant is related to the ratio of heat and power demands we should again remember that the absolute size of those demands is also critical. For example, a single Diesel engine has a relatively low power output, and although several engines may be run in parallel to meet a large power demand, gas or steam turbines are likely to be used to meet higher powers (see the examples in Chapter 5).

3.8   Design of CHP Plant for Varying Plant Heat to Power Ratio (λ_{CG})

The calculation of the last section shows how for the three plants considered—back-pressure turbine, gas turbine and Diesel engine with waste heat recuperators—both the plant heat to power ratio (λ_{CG}) and performance criteria (such as FESR) are affected by choice of design parameters

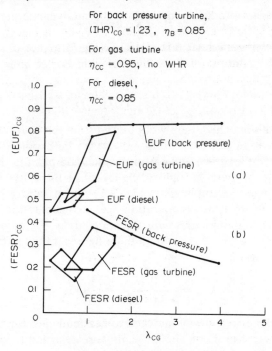

For back pressure turbine,
$(IHR)_{CG} = 1.23$, $\eta_B = 0.85$

For gas turbine
$\eta_{CC} = 0.95$, no WHR

For diesel,
$\eta_{CC} = 0.85$

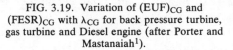

FIG. 3.19. Variation of $(EUF)_{CG}$ and
$(FESR)_{CG}$ with λ_{CG} for back pressure turbine,
gas turbine and Diesel engine (after Porter and
Mastanaiah[1]).

within the CHP plant itself (e.g. steam turbine inlet conditions, gas turbine pressure ratio, etc.). However, the range of λ_{CG} covered by each type of plant is not large, as Fig. 3.19 indicates, and operation may have to be in parallel with "supplementary" plant, as was described in section 3.4, if λ_D is substantially different from λ_{CG}.

There are, however, two plants where some range of λ_{CG} can be achieved by "internal" design, and matched operation with λ_{CG} equal to (varying) λ_D may prove possible. These plants are (i) a gas turbine with or without a heat exchanger, and (ii) an extraction steam turbine, for which the degree of extraction is to some extent a (thermodynamic) design variable. We consider first the gas turbine case, in general fashion, but study the extraction steam turbine in more detail, particularly the interaction between extraction of steam for feed heating of the water flow returning direct to the boiler, and extraction of steam for district heating. In large plant (see Oliker[8]) steam is extracted for both these purposes.

3.8.1   Gas Turbine with Waste Heat Recuperator

The analysis of section 3.4.3 showed how performance of a gas turbine in parallel with a conventional plant varied with λ_D. For large values of λ_D

$$\left(\lambda_D > \lambda_{CG} = \frac{(1 - \eta_{CG})}{\eta_{CG}[1 + (Q_{NU}/Q_U)_{CG}]}\right),$$

if the heat load is matched, then work output from the conventional power plant becomes negative, i.e. electric power can be exported to the grid; or if the electrical load is matched, surplus heating plant is required.

Timmermans[9] suggests a high efficiency (0.3) for a gas turbine with no heat exchanger (example C of Fig. 2.1) and a ratio $(Q_{NU}/Q_U)_{CG}$ of 0.193 for the waste heat recuperation (i.e. a recuperation "efficiency" of about 0.84), although this will be affected by limitations on stack temperature (see section 3.4.3). With these figures it follows that the plant heat to power ratio is

$$\lambda_{CG} = \frac{0.7}{0.3 \cdot 1.193} = 1.955.$$

Such a gas turbine is thus well matched to heat and electricity demands of ratio $\lambda_D \approx 2$. Suppose that a similar plant were designed to meet a winter value of $\lambda_D = 2$ (Fig. 3.20(a)); and that in summer the electricity load

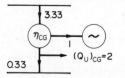

(a) Plant matched to winter demands ($\lambda_D = 2$)

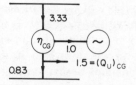

(b) Summer demands ($\lambda_D = 1.5$) more reject heat wasted

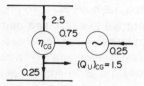

(c) Summer demands ($\lambda_D = 1.5$) extra power from grid

FIG. 3.20. Off design performance of CHP plant.

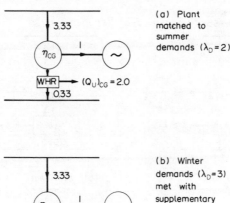

(a) Plant matched to summer demands ($\lambda_D = 2$)

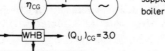

(b) Winter demands ($\lambda_D = 3$) met with supplementary boiler

FIG. 3.21. Off design performance of CHP plant.

remained constant while the heat load dropped to say 75% of the winter load (so that $\lambda_D = 1.5$). Assuming that the plant itself still operated at about the same value of $\lambda_{CG} = 2$, then either it could continue to be run at full load but with an increase in wasted heat (Fig. 3.20(b)) or it could be run at part load (Fig. 3.20(c)) with 25% "make-up" electricity from the grid.

Alternatively, if the demand values of λ_D were 3 (for the winter) and 2 (for the summer), then the plant could be designed for the summer loads (Fig. 3.21(a)). In this case, if the electrical demand were constant for summer to winter, heat would be required from a waste heat boiler (HRSG fired) in winter to meet the extra heat load (see Fig. 3.21(b)).

These are simplified arguments, for the overall selection of plant size and type will require detailed study of the daily load variations as well as the discussed variations, and assessment of the economies of operation. The performance of CHP plant "off-design" to meet diurnal variations is discussed later in Chapter 4, but here we are concerned with a gross change in λ_D from summer to winter.

An interesting proposal to deal with the different values of λ_D demanded in summer and winter has been put forward by Lowder in several papers (see for example Lowder[10]). Lowder considers a gas turbine plant with the following characteristics:

$$\eta_{CG} = 0.172 \text{ (no heat exchanger)}$$
$$\eta_{CG} = 0.235 \text{ (with a heat exchanger of thermal ratio 0.8)}$$

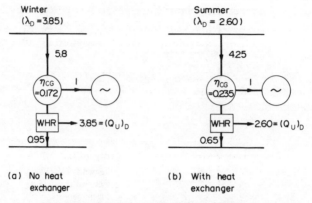

FIG. 3.22. Off design performance of CHP plant.

If it is assumed that the "efficiency" of recuperation of a waste heat boiler is also 0.8, then the basic plant *without* a heat exchanger would yield a value of

$$\lambda_{CG} = \frac{0.828}{0.172 \cdot 1.25} = 3.85.$$

If the plant operates *with* a heat exchanger, and if a waste heat recuperator of "efficiency" 0.8 is used downstream of the heat exchanger then for this plant

$$\lambda_{CG} = \frac{0.765}{0.235 \cdot 1.25} = 2.60.$$

Thus a plant which operates without a heat exchanger may be used to match winter heat and electricity loads (high λ_D). In summer the heat exchanger may be "switched in" and meet the requirements of a lower value of λ_D (constant electricity demand, lower heat load) (see Fig. 3.22).

Such arguments are simplistic, but illustrate an important concept, a useful flexibility of the gas turbine to meet different values of λ_D. In practice, as Lowder points out, exhaust stack temperature of the gas turbine is limited to a minimum value (by the necessity to avoid condensation) and this controls the ratio of non-useful heat rejection to useful heat supply $(Q_{NU}/Q_U)_{CG}$.

3.8.2   The Steam Extraction Plant

The general analysis of section 3.4.2 for a steam extraction cycle has validity for both reversible CHP plant (the reversible Carnot engines of section 3.6.1) and irreversible plant. Care has to be exercised in definitions of the topping and bottoming efficiencies (η_H, η_L), but we can amplify the analysis to show how the plant may be designed to meet different values of λ_D.

We consider first a steam plant with single extraction for district (or process) heating, relating the general analysis of section 3.4.2 to the fraction of steam flow that is bled.

However, a large steam plant with steam extraction for district heating will almost certainly also use bled steam for feed heating (to increase basic efficiency by raising the mean temperature of heat supply). In the low pressure turbine steam is bled for both purposes, supplying both feed heating and district heating. We discuss this problem using a simple analysis, based on the assumption of a particular approximation to the turbine condition line (Horlock[11]).

3.8.2.1 Single Extraction of Steam for District Heating (Relation of Mass Extracted (m) to λ_{CG})

It is first necessary to explain how the general description of Fig. 3.4 (in which heat was (partially) rejected from an upper to a lower cycle) has relevance to a CHP plant with pass-out steam. Figure 3.23(a) gives an interpretation of the earlier generalised analysis, showing how steam extraction may be incorporated within that analysis. It is supposed that unit quantity of the working fluid circulates in the upper plant, and $(1 - m)$ of the same working fluid in the lower plant. The quantities originally shown in Fig. 3.4 are reproduced in Fig. 3.23(a), and related to the (specific) "working" enthalpy drops in the upper and lower cycles Δh_H and Δh_L, (quantities per unit mass denoted by lower case symbols).

It is now assumed that the fluid in the upper cycle is split after heat rejection q_{HL} into two streams $(1 - m)$ and m. The first stream is reheated through the same temperature levels, and the second is not. Heat $(mq_{HL}) = \lambda_{CG}$ is thus available to meet the heat load λ_D. Essentially m is therefore the bled steam extracted, and this quantity can be related to the two component efficiencies, η_H and η_L, and the heat load $\lambda_D = \lambda_{CG}$.

The ratio λ_{CG} is

$$\lambda_{CG} = \frac{Q_U}{W_H + W_L} = \frac{mq_{HL}}{\Delta h_H + (1 - m)\Delta h_L}. \qquad (3.106)$$

But

$$W_H = \Delta h_H = \eta_H f = \eta_H(q_{HL} + \Delta h_H), \qquad (3.107)$$

so that

$$\frac{\Delta h_H}{q_{HL}} = \frac{\eta_H}{(1 - \eta_H)}. \qquad (3.108)$$

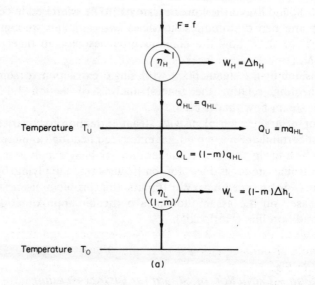

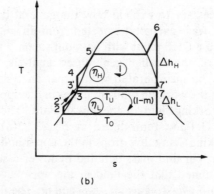

FIG. 3.23. Extraction or pass-out turbine.

Also

$$\eta_L = \frac{(1 - m)\Delta h_L}{(1 - m)q_{HL}} = \frac{\Delta h_L}{q_{HL}}. \tag{3.109}$$

Substitution in equation (3.106) yields

$$m = \frac{\lambda_{CG}(\eta_H + \eta_L - \eta_H\eta_L)}{(1 + \lambda_{CG}\eta_L)(1 - \eta_H)}. \tag{3.110}$$

The various quantities (Δh_H, Δh_L etc.) are illustrated in Fig. 3.23(b) for a Rankine-type cycle with condensation of pass-out steam. The upper cycle

(H) is $2'3'4567'3'2'$ and the lower cycle is 123781 where states indicated by a dash are virtually identical to those without. Feed-pump work is neglected, so Δh_H, Δh_L are the enthalpy drops in the H.P. and L.P. cylinders respectively. (If feed pump-work is included then the pump-work must be subtracted from the turbine enthalpy drops to give the "working" enthalpy drops.) Further, $q_{HL} = h_7 - h_2$, $f = h_6 - h_2$, $\Delta h_H = h_6 - h_7$, and $\Delta h_L = h_7 - h_8$.

These quantities are straightforward in their definition, but it should be noted that the upper efficiency η_H is equal to $\Delta h_H/(h_6 - h_2)$ and η_L is equal to $\Delta h_L/(h_7 - h_2)$.

If the efficiencies defined on this basis were $\eta_H = 0.33$ and $\eta_L = 0.1$, then $\eta_H\eta_L = 0.033$ and the overall Rankine efficiency would be

$$\eta_H + \eta_L - \eta_H\eta_L \approx 0.4,$$

the figure used in section 3.8.3. The performance parameters (EUF)', (FESR)' were given in equations (3.95) and (3.96), and they were plotted against λ_D in Fig. 3.12.

Further, we can now express m in terms of λ_{CG}, from equation (3.110), and it is also shown plotted in Fig. 3.12.

The plant operates as a back-pressure set for a heat to electricity demand ratio of $\lambda_{CG} = 2.0$ with $m = 1.0$ (all the steam is "bled" at T_U). However, it can be designed for smaller values of $\lambda_{CG} = \lambda_D$, with less extraction. For example, at $\lambda_{CG} = 1.0$, $m = 0.54$ when $(EUF)_{CG} = 0.73$ and $(FESR)_{CG} = 0.24$; at $\lambda_{CG} = 0.25$, $m = 0.286$, $(EUF)_{CG} = 0.57$ and $(FESR)_{CG} = 0.14$.

For a particular plant, designed for a specified λ_D, varying steam extraction may be difficult to achieve because of limitations on the L.P. turbine performance. But some variation in the extraction fraction m enables the plant to operate at variable λ, giving some flexibility for summer and winter operation.

3.8.2.2   Extraction for Feed Heating and District Heating

As explained earlier, in a big steam turbine plant steam may be extracted for both feed heating and district heating. Figure 3.24 shows this practice diagrammatically; in general, extraction for both feed heating and district heating will be undertaken at the low pressures, but extraction for feed heating alone will be taken at higher pressure levels.

A simple analysis of this cycle has been given by Horlock[11], based on a particular assumption for the turbine expansion line—that the difference between local steam enthalpy (h) and the enthalpy of water at the same pressure (h_f) is constant (equal to β). This approximation, illustrated in Fig. 3.25, was originally made by Salisbury[12].

That this is quite a good assumption may be illustrated by modification

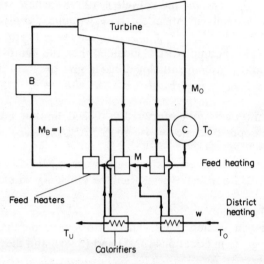

FIG. 3.24. Combined district and feed
heating.

of the Rankine cycle considered in section 2.6.2, to allow the steam
expansion in the turbine to take place according to the condition that
β = constant, rather than along an isentropic. The entry conditions to the
turbine were 2 MPa and 350°C, giving an enthalpy of 3138.6 kJ/kg, and
for isentropic expansion to the condenser pressure of 7 kPa (39°C), the

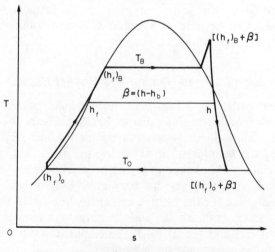

FIG. 3.25. Basic cycle showing "constant β"
turbine expansion line.

enthalpy at exit was 2161.4 kJ/kg, giving an enthalpy drop of 977.2 kJ/kg. At entry, $\beta = 3138.6 - 908.6 = 2230.0$. For an irreversible expansion β = constant, the enthalpy at turbine exit will be $(163.4 + 2230.0) = 2393.4$ kJ/kg giving an enthalpy drop, $= 3138.6 - 2393.4 = 745.2$ kJ/kg. The turbine efficiency is thus $(745.2/977.2) = 76.26\%$ which is reasonably realistic, if a little low. It is of interest to note that the enthalpy drop in the turbine is equal to the difference between the enthalpy of saturated water at the boiler pressure $(h_f)_B$ and the enthalpy of saturated water at the condenser pressure $(h_f)_0$. (The enthalpy drop in the turbine is

$$\alpha = [(h_f)_B + \beta] - [(h_f)_0 + \beta]$$
$$= [(h_f)_B - (h_f)_0].) \tag{3.111}$$

Using the assumption β = constant, we derive simple expressions for the change in performance of the plant if steam is bled for district heating alone; and for district heating and feed heating. We consider the following cycles.

(i) Basic Rankine cycle (no feed heating or district heating).
(ii) Cycle with single bleed for district heating to useful temperature T_U.
(iii) Cycle with "continuous" bleed for a large number of feed heaters and district heaters to useful temperature T_U, plus continuous bleed for further feed heating only to boiler temperature T_B.
(A more complete development of this approach is given by Horlock[11].)

To simplify the analyses (and the subsequent numerical calculations) we assume that the temperature T_U (for district heating) is the arithmetic mean of the temperatures T_0 and T_B. (This is consistent with Haywood's

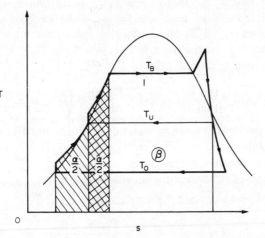

FIG. 3.26. Areas α and β.

recommendation for a single bleed if the specific heat of water is reasonably constant; he suggests that the enthalpy rise of the feed water (α) is divided equally between the various heaters used (including the economiser).) Figure 3.26 shows the two areas $\dfrac{\alpha}{2}$, and the area β on a T,s diagram.

(i) Basic Cycle

For the basic cycle the cycle efficiency is simply

$$
\begin{aligned}
\eta &= \frac{\text{Work Output}}{\text{Heat Supplied}} \\[2mm]
&= \frac{\text{Heat Supplied} - \text{Heat Rejected}}{\text{Heat Supplied}} \\[2mm]
&= \frac{(\alpha + \beta) - \beta}{(\alpha + \beta)} \\[2mm]
&= \frac{\alpha}{\alpha + \beta}.
\end{aligned}
\tag{3.112}
$$

(ii) Cycle with Single Bleed for District Heating

For the cycle with a single extraction point, the bled steam (m) raises the district heating water (mass flow w) from temperature T_0 to the useful temperature T_U in a calorifier.

We consider two cases. In the first (Fig. 3.27) the bled steam yields heat both as it condenses and also subsequently as water, dropping in temperature from T_U to T_0 (the calorifier must be contra flow without heat loss) before it joins the main condensate flow, $M_0 = (1 - m_0)$.

The heat balance in the calorifier is then

$$
Q_U = wc_p(T_U - T_0) = \frac{w\alpha}{2} = m\left(\beta + \frac{\alpha}{2}\right),
\tag{3.113}
$$

assuming that the district heating water enters the calorifier at condensate temperature (T_0).

The total heat rejected is the sum of the useful heat (Q_U) and the non-useful heat rejected in the main condenser (Q_{NU}),

$$
Q_A = m\left(\beta + \frac{\alpha}{2}\right) + (1 - m)\beta = \beta + \frac{m\alpha}{2},
\tag{3.114}
$$

and the heat supplied is

$$
Q_B = \alpha + \beta.
\tag{3.115}
$$

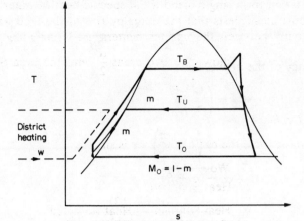

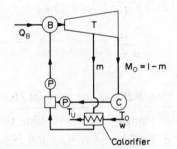

FIG. 3.27. Cycle with single bleed for district
heating (boiler heats feed water from T_0).
After Horlock.[11]

Thus the work output is

$$W = (\alpha + \beta) - \left(\beta + \frac{m\alpha}{2}\right)$$

$$= \alpha\left(1 - \frac{m}{2}\right), \tag{3.116}$$

and the useful heat to work ratio is

$$\lambda_{CG} = \left(\frac{Q_U}{W}\right)_{CG} = \frac{m\left(\beta + \dfrac{\alpha}{2}\right)}{\alpha\left(1 - \dfrac{m}{2}\right)}, \tag{3.117}$$

so that

$$m = \frac{2\lambda_{CG}}{\alpha(1 + \lambda_{CG}) + 2\beta}.$$ (3.118)

The efficiency of the new cycle is

$$\eta_{CG} = \left(\frac{W}{Q_B}\right)_{CG}$$

$$= \frac{\alpha\left(1 - \frac{m}{2}\right)}{(\alpha + \beta)}$$

$$= \frac{\alpha(\alpha + 2\beta)}{(\alpha + \beta)[(\lambda_{CG} + 1)\alpha + 2\beta]},$$ (3.119)

and the energy utilisation factor is

$$(EUF)_{CG} = (\lambda_{CG} + 1)\eta_{CG}$$

$$= \frac{(\lambda_{CG} + 1)\alpha(\alpha + 2\beta)}{(\alpha + \beta)[(\lambda_{CG} + 1)\alpha + 2\beta]}.$$ (3.120)

We may note that the "Z-factor" is

$$Z = \frac{\text{Lost Work}}{\text{Useful Heat}} = \frac{m\alpha/2}{m[\beta + (\alpha/2)]} = \left(\frac{\alpha}{\alpha + 2\beta}\right).$$ (3.121)

This analysis can be compared with that of section 3.8.2.2, which was based on estimated efficiencies of the "upper" and "lower" cycles, η_H and η_L. These two efficiencies, expressed in terms of α and β, are

$$\eta_H = \frac{(\alpha + \beta) - [\beta + (\alpha/2)]}{(\alpha + \beta)} = \frac{\alpha}{2(\alpha + \beta)},$$ (3.122)

$$\eta_L = \frac{[\beta + (\alpha/2)] - \beta}{\left[\beta + \left(\frac{\alpha}{2}\right)\right]} = \frac{\alpha}{\alpha + 2\beta} = Z,$$ (3.123)

so that the "conventional" plant efficiency is

$$\eta_C = \eta_H + \eta_L - \eta_H\eta_L$$

$$= \frac{\alpha}{\alpha + \beta},$$ (3.124)

as given above, for the basic cycle (i).

Equation (3.110) for the mass flow extracted in a cycle with single bleed was expressed in terms of η_C, η_H and η_L as

$$m = \frac{\lambda_{CG}\eta_C}{(1 + \lambda_{CG}\eta_L)(1 - \eta_H)}$$

$$= \frac{\dfrac{\lambda_{CG}\alpha}{(\alpha + \beta)}}{\left[1 + \left(\dfrac{\lambda_{CG}\alpha}{\alpha + 2\beta}\right)\right]\left[1 - \dfrac{\alpha}{2(\alpha + \beta)}\right]}$$

$$= \frac{2\lambda_{CG}\alpha}{(\lambda_{CG} + 1)\alpha + 2\beta}, \tag{3.125}$$

which is equation (3.118), and which yields equation (3.119) and (3.120) for $(\eta)_{CG}$ and $(EUF)_{CG}$.

The second approach (Fig. 3.28) involves the assumption that the bled steam gives up heat only in condensation before it rejoins the main con-

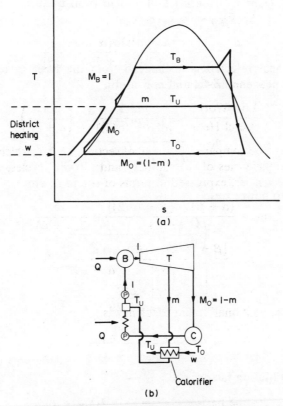

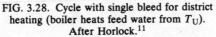

FIG. 3.28. Cycle with single bleed for district heating (boiler heats feed water from T_U). After Horlock.[11]

densate flow (M_0) at temperature T_U. Then the heat balance in the calorifier is

$$Q_U = wc_p(T_U - T_0) = \frac{w\alpha}{2} = m\beta. \qquad (3.126)$$

The efficiency of the plant as a cogeneration scheme is now

$$\eta_{CG} = \frac{\text{Heat Supplied} - \text{Heat Rejected}}{\text{Heat Supplied}},$$

where the heat supplied

$$Q_B = (1 - m)\frac{\alpha}{2} + \frac{\alpha}{2} + \beta,$$

and the heat rejected

$$\begin{aligned} Q_A &= Q_U \text{ (useful)} + (1 - m)\beta \text{ (non-useful)} \\ &= m\beta + (1 - m)\beta \\ &= \beta. \end{aligned}$$

(The overall heat rejected is unchanged from the basic cycle, case (i), but is divided between useful and non-useful.)

Thus
$$\eta_{CG} = \frac{(1 - m)\dfrac{\alpha}{2} + \dfrac{\alpha}{2}}{(1 - m)\dfrac{\alpha}{2} + \dfrac{\alpha}{2} + \beta}. \qquad (3.127)$$

But the heat to power ratio

$$\lambda_{CG} = \left(\frac{Q_U}{W}\right)_{CG} = \frac{m\beta}{(1 - m)\dfrac{\alpha}{2} + \dfrac{\alpha}{2}}, \qquad (3.128)$$

so that
$$m = \frac{2\lambda_{CG}\alpha}{\lambda_{CG}\alpha + 2\beta}, \qquad (3.129)$$

and
$$\eta_{CG} = \frac{2\alpha}{2(\alpha + \beta) + \lambda_{CG}\alpha}. \qquad (3.130)$$

The energy utilisation factor is

$$(\text{EUF})_{CG} = (\lambda_{CG} + 1)\eta_{CG} = \frac{2(\lambda_{CG} + 1)\alpha}{2(\alpha + \beta) + \lambda_{CG}\alpha}. \qquad (3.131)$$

(The "Z-factor" in this case is

$$Z = \frac{\text{Lost Work}}{\text{Useful Heat}} = \frac{m\alpha/2}{m\beta} = \frac{\alpha}{2\beta}.)$$ (3.132)

(iii)   Cycle with "Continuous Bleed" (for District and Feed Heating)

We next consider a cycle with steam extracted "continuously" from the turbine to provide both feed and district heating in an infinite number of heaters (Fig. 3.29). We now define $m(T)$ as the mass of steam bled between T_0 and T, and $M(T)$ as the mass of feed water flowing into an

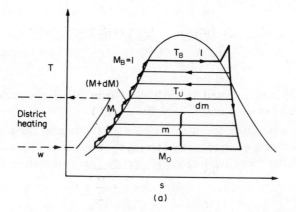

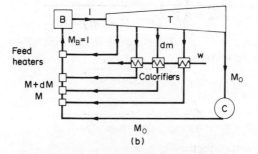

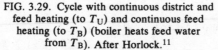

FIG. 3.29. Cycle with continuous district and feed heating (to T_U) and continuous feed heating (to T_B) (boiler heats feed water from T_B). After Horlock.[11]

"elementary" (direct contact) feed heater at temperature T. There is unit flow through the boiler.

We assume that "continuous" district *and* feed heating takes place between T_0 and T_U, with subsequent feed heating only, from T_U to T_B. The total amount of steam extracted is $m_B = \int_{T_0}^{T_B} dm$ (m_0 is zero by definition). The mass flow through the condenser is $M_0 = 1 - m_B$ and at any temperature T, $M(T) = M_0 + m(T)$.

Between temperatures T_0 and T_U, with both feed *and* district heating, we may write the heat balance for an elementary quantity of bled steam (dm), transferring heat $\beta\, dm$ to *both* the district heating water flow (w) *and* the feed water flow (M), as

$$\beta\, dm = \beta\, dM = (w + M)c_p\, dT.$$

so that

$$(w + M) = C_1 e^{(c_p T/\beta)}. \tag{3.133}$$

where β is the specific enthalpy change introduced earlier. Between temperatures T_U and T_B, *with feed heating only*, the heat balance is

$$\beta\, dm = \beta\, dM = Mc_p\, dT,$$

so that

$$M = C_2 e^{(c_p T/\beta)}. \tag{3.134}$$

The mass flow through the main condenser (at temperature T_0) is M_0, and if the mass flow of condensate leaving the "combined" heating section and entering the "feed-heating only" section (at temperature T_U) is M_U, then at T_U,

$$(M_U + w) = C_1 e^{(c_p T_U/\beta)}, \tag{3.135}$$

and at T_0,

$$(M_0 + w) = C_1 e^{(c_p T_0/\beta)}. \tag{3.136}$$

Hence

$$\frac{(M_U + w)}{(M_0 + w)} = e^{c_p(T_U - T_0)/\beta} = e^{\alpha/2\beta}, \tag{3.137}$$

if the temperature rise $(T_U - T_0)$ is equal to the rise $(T_B - T_U)$. For the "feed heating" only section,

at T_U,

$$M_U = C_2 e^{(c_p T_U/\beta)},$$

and at T_B

$$M_B = 1 = C_2 e^{(c_p T_B/\beta)},$$

so that

$$M_U = e^{c_p(T_U - T_B)/\beta}$$
$$= e^{-\alpha/2\beta}, \tag{3.138}$$

Equations (3.137) and (3.138) yield

$$w = (M_0 - e^{-\alpha/\beta})/(e^{-\alpha/2\beta} - 1). \tag{3.139}$$

The useful heat rejection is

$$Q_U = wc_p(T_U - T_0) = \frac{(M_0 - e^{-\alpha/\beta})\alpha}{2(e^{-\alpha/2\beta} - 1)}, \tag{3.140}$$

so that

$$M_0 = \frac{2Q_U}{\alpha}(e^{-\alpha/2\beta} - 1) + e^{-\alpha/\beta}. \tag{3.141}$$

The efficiency of the cycle is

$$\eta_{CG} = 1 - \frac{(M_0\beta + Q_U)}{\beta}$$

$$= 1 - e^{-\alpha/\beta} - \frac{Q_U}{\beta}\left[1 + \frac{2\beta}{\alpha}(e^{-\alpha/2\beta} - 1)\right]. \tag{3.142}$$

But the heat to power ratio is

$$\lambda_{CG} = \frac{Q_U}{\beta(1 - M_0) - Q_U}, \tag{3.143}$$

so that

$$Q_U = \frac{\lambda_{CG}\beta(1 - e^{-\alpha/\beta})}{(1 + \lambda_{CG}) + \frac{2\lambda\beta}{\alpha}(e^{-\alpha/2\beta} - 1)}. \tag{3.144}$$

It then follows that the efficiency is

$$\eta_{CG} = \frac{1 - e^{-\alpha/\beta}}{(\lambda_{CG} + 1) + \frac{2\lambda_{CG}\beta}{\alpha}(e^{-\alpha/2\beta} - 1)}, \tag{3.145}$$

and the energy utilisation factor is

$$(EUF)_{CG} = \frac{(\lambda_{CG} + 1)(1 - e^{-\alpha/\beta})}{(\lambda_{CG} + 1) + \frac{2\lambda_{CG}\beta}{\alpha}(e^{-\alpha/2\beta} - 1)}. \tag{3.146}$$

If there is no district heating (no useful heat (Q_U) rejected, $\lambda_{CG} = w = 0$) then

$$\eta_C = (EUF) = 1 - e^{-\alpha/\beta}, \tag{3.147}$$

which is the efficiency of a cycle with "continuous" feed heating between T_0 and T_B.

Two other special cases are discussed in detail by Horlock[11], but only the results are given here.

(a) If feed heating stops at T_U (in a cycle with feed *and* district heating between T_0 and T_U) then it may be shown that

$$\eta_{CG} = \frac{1 - e^{-\alpha/2\beta} + (\alpha/2\beta)}{[1 + (\alpha/2\beta)]\left[1 + \lambda_{CG} - \dfrac{2\lambda_{CG}\beta}{\alpha}(1 - e^{-\alpha/2\beta})\right]}, \quad (3.148)$$

$$(\text{EUF})_{CG} = \frac{(\lambda_{CG} + 1)\left[\dfrac{\alpha}{2\beta} + (1 - e^{-\alpha/2\beta})\right]}{[1 + (\alpha/2\beta)]\left[1 + \lambda_{CG} - \dfrac{2\lambda_{CG}\beta}{\alpha}(1 - e^{-\alpha/2\beta})\right]}. \quad (3.149)$$

(b) If there is no feed heating, but "continuous" district heating between T_0 and T_U then

$$\eta_{CG} = \frac{(\alpha/\beta)}{[1 + (\alpha/\beta)] + (\lambda_{CG}\alpha/4\beta)}, \quad (3.150)$$

$$(\text{EUF})_{CG} = \frac{(1 + \lambda)(\alpha/\beta)}{[1 + (\alpha/\beta)] + (\lambda_{CG}\alpha/4\beta)}. \quad (3.151)$$

Calculations for the modifications (ii) and (iii) to the basic cycle (i) are illustrated in Fig. 3.30. In the basic cycle the turbine received steam from the boiler at 2 MPa and 350°C, and exhausted to the condenser at 7 MPa. Values of α and β were 745.2 kJ/kg and 2230.0 (kJ/kg) respectively, and the basic efficiency was

$$\eta_C = \frac{\alpha/\beta}{1 + (\alpha/\beta)} = \frac{0.3342}{1.3342} = 0.2505.$$

(The Rankine cycle efficiency, ignoring feed pump terms, for these conditions is given by Haywood[13] as 0.3284. With a turbine efficiency (for a "constant" expansion) of 0.7625, the overall efficiency becomes $0.7625 \times 0.3284 = 0.2505$, neglecting feed pump work.)

Single bleed district heating to T_U (cycle (ii)) lowers this efficiency, but (EUF) increases with λ_{CG} (and the mass of bled steam). There is a limiting value of (about 6) when *all* the steam is bled, and the plant becomes a back-pressure turbine.

Calculations for cycle (iii) (continuous feed heating *and* district heating to T_U, and subsequent feed heating only to T_B) show that this cycle achieves the best performance (Fig. 3.30). However, the advantages of "continuous" heating are not enormously large; this conclusion is similar to that reached for feed heating only many years ago—that a finite number of heaters may achieve a substantial fraction of the "ultimate" advantage of feed heating by a continuous number of heaters (see Haywood[13]).

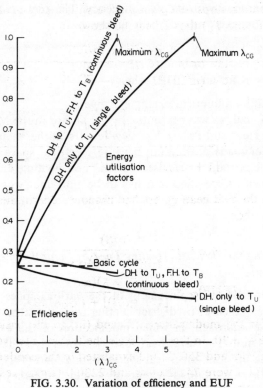

FIG. 3.30. Variation of efficiency and EUF
with λ (heat to power ratio).
After Horlock.[11]

3.9   Fuel Savings from Installation of CHP Plant for a District Heating Scheme

The previous sections in this chapter have given derivations of the FESR (fuel energy savings ratio) for several types of CHP plant, together with some simple numerical calculations. In each case the FESR was obtained by comparison with a reference case—separate power and boiler plants meeting the electricity and heat loads.

In this section, we repeat this type of comparative analysis but in more detailed fashion for a district heating scheme. In particular the "basic" heat load (that of a hypothetical "average" house) is supposed to be met by an existing mix of gas heating, coal heating and electrical (Joule) heating. Thus instead of the single boiler of the "basic" system of section 3.2 (meeting a heat load), there are several sources of heat, each meeting part of the heat load with its own efficiency. The fuel savings associated with change from this "mixed" heat supply to

(i) a single "heat only" boiler;
(ii) a back-pressure turbine (CHP) scheme;
(iii) a pass-out turbine (CHP) scheme

are assessed and a numerical example is given.

This type of analysis was presented by the author in the Marshall report on Combined Heat and Power[14]. A new feature in the analysis is that the heat and power loads are linked, because electrical space heating is included. It is important to realise that the non-heating electrical power demand is taken as specified and has to be fully met when the plant is changed from the reference case (the "change" type of analysis referred to earlier).

3.9.1   Supplying the Domestic Heating Load by Central Boiler Plant

We consider first the replacement of the various space heating loads (now the reference case) by district heating from a central boiler; Figure 3.31 shows the approach, in which modification of the existing heating plant is considered. The total heat load $(\lambda)^*$ of a hypothetical "average house" is made up of an electrical space heating load $(\dot{W}_S)$, and domestic heating loads $(\dot{\lambda}_1, \dot{\lambda}_2)$ supplied by units of "boiler" efficiency η_{B_1}, η_{B_2}. These are all replaced by district heating from the central boiler of efficiency $(\eta_B)'$, and the heat demand changes to $(\dot{\lambda})'$ (the domestic "comfort" is increased). The fuel energy supplied to the central boiler is $(\dot{\lambda})'(1 + \phi)/(\eta_B)'$ compared with the fuel energy $[(\dot{\lambda}_1/\eta_{B_1}) + \dot{\lambda}_2/\eta_{B_2}]$ supplied in the domestic heating plants.

Originally, the grid electrical power output met both the electrical demand for space heating $(\dot{W}_S)$, as well as some basic power demand (of unity). The space heating load is now taken by the central boiler, so that the electrical output from the conventional plant can effectively be reduced, leading to an energy saving on the fuel originally supplied to the "grid" power plants. (It should be noted that although the external power load is taken as unity, it could have any arbitrary value, since in calculating the fuel savings in changing from conventional to CHP plant, the external load is assumed constant and eliminated in the analysis. $\dot{\lambda}$ does not therefore have a significance as the heat/power ratio in this study of district heating).

---

*Loads are now defined as energy *rates* $\dot{\lambda}, \dot{W}$, dot superscripts being introduced (except for the power demand which is arbitrarily assumed to be unity). This is to enable the analysis to be used in Chapter 4, where the economics of CHP plant are related to such energy rates.

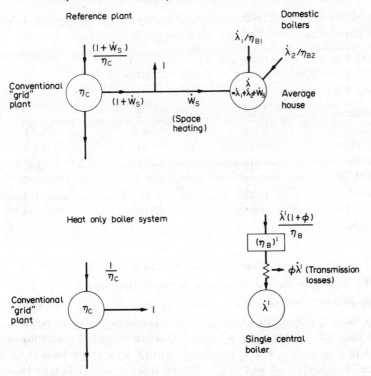

FIG. 3.31. Fuel savings in district heating by heat only boilers.

Thus the total energy saving is

$$\Delta \dot{F}_{a} = \dot{F}_{REF} - \dot{F}_{CG} = \left[\frac{\lambda_1}{\eta_{B_1}} + \frac{\lambda_2}{\eta_{B_2}} + \frac{(1 + \dot{W}_S)}{\eta_C} \right] - \left[\frac{(\dot{\lambda})'(1 + \phi)}{(\eta_B)'} + \frac{1}{\eta_C} \right],$$
(3.152)

where ϕ is the loss factor allowing for transmission losses (Fig. 3.31) and η_C is the overall efficiency of the "external" conventional plant. $\Delta \dot{F}_a$ may be expressed in non-dimensional form, by dividing by the original heat load $\dot{\lambda} = (\dot{\lambda}_1 + \dot{\lambda}_2 + \dot{W}_S)$,

$$\left(\frac{\Delta \dot{F}}{\dot{\lambda}} \right)_a = \frac{\dot{\lambda}_1}{\dot{\lambda} \eta_{B_1}} + \frac{\dot{\lambda}_2}{\dot{\lambda} \eta_{B_2}} + \frac{\dot{W}_S}{\dot{\lambda} \eta_C} - \frac{(\dot{\lambda})'(1 + \phi)}{\dot{\lambda}(\eta_B)'},$$
(3.153a)

or $\qquad \gamma_a = \left(\frac{\Delta \dot{F}}{\dot{\lambda}} \right)_a = \gamma_1 + \gamma_2 + \gamma_S - (\gamma_B)',$ (3.153b) where $$\gamma_1 = \frac{\dot{\lambda}_1}{\dot{\lambda}\eta_{B_1}}, \qquad \gamma_2 = \frac{\dot{\lambda}_2}{\dot{\lambda}\eta_{B_2}}, \qquad \gamma_S = \frac{\dot{W}_S}{\dot{\lambda}\eta_C},$$

$$(\gamma_B)' = \frac{(\dot{\lambda})'}{\dot{\lambda}} \frac{(1 + \phi)}{(\eta_B)'} = \frac{\delta(1 + \phi)}{(\eta_B)'}, \qquad \delta = \frac{(\dot{\lambda})'}{\dot{\lambda}}.$$

An alternative form is the fuel energy savings ratio, expressed in terms of the energy originally required, rather than the original heat load,

$$(\text{FESR})'_a = \frac{(\Delta\dot{F}/\dot{\lambda})_a}{\left[\dfrac{\dot{\lambda}_1}{\dot{\lambda}\eta_{B_1}} + \dfrac{\dot{\lambda}_2}{\dot{\lambda}\eta_{B_2}} + \dfrac{(1 + \dot{W}_S)}{\dot{\lambda}\eta_C}\right]} = 1 - \frac{(\gamma_B)' + \left(\dfrac{1}{\dot{\lambda}\eta_C}\right)}{\gamma_1 + \gamma_2 + \gamma_S + \left(\dfrac{1}{\dot{\lambda}\eta_C}\right)}.$$

$$(3.154)$$

3.9.2 Supplying the Electrical and Domestic Heating by a Back-Pressure Turbine

We next consider replacement of the existing mix of heating by rejected heat from a back-pressure turbine. Figure 3.32 shows the basic plant as described in section 3.9.1, together with the new CHP plant. The back-pressure turbine plant has an overall efficiency of η_{CG} (less than that of the "grid" turbine plant) and receives fuel energy $\dfrac{\dot{W}_{CG}}{\eta_{CG}} = \dfrac{(1 - \dot{W}_C)}{\eta_{CG}}$. Its electrical power output $(1 - \dot{W}_C)$ goes to meet only part of the original (non-heating) electrical power load of unity; its efficiency is less than that of the "grid" turbine, and power has to be supplied from the grid of amount $\dot{W}_C$. The total fuel energy required from the back-pressure plant and the grid turbine is therefore $\left[\dfrac{\dot{W}_{CG}}{\eta_{CG}} + \dfrac{\dot{W}_C}{\eta_C}\right] = \left[\dfrac{(1 - \dot{W}_C)}{\eta_{CG}} + \dfrac{\dot{W}_C}{\eta_C}\right]$. The back-pressure set meets the full heat load, and no extra boilers are required.

The fuel savings (compared with the reference case) are therefore

$$\Delta\dot{F}_b = \dot{F}_{REF} - \dot{F}_{CG}$$

$$= \left[\frac{\dot{\lambda}_1}{\eta_{B_1}} + \frac{\dot{\lambda}_2}{\eta_{B_2}} + \frac{(1 + \dot{W}_S)}{\eta_C}\right] - \left[\frac{(1 - \dot{W}_C)}{\eta_{CG}} + \frac{\dot{W}_C}{\eta_C}\right]$$

$$= \left[\frac{\dot{\lambda}_1}{\eta_{B_1}} + \frac{\dot{\lambda}_2}{\eta_{B_2}} + \frac{\dot{W}_S}{\eta_C}\right] - (1 - \dot{W}_C)\left(\frac{1}{\eta_{CG}} - \frac{1}{\eta_C}\right). \qquad (3.155)$$

Reference plant

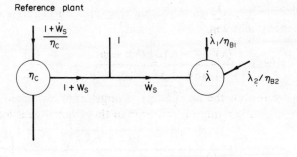

Back pressure turbine system

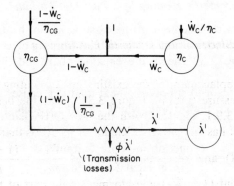

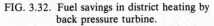

FIG. 3.32. Fuel savings in district heating by
back pressure turbine.

But if there are losses $\phi(\lambda)'$ in supplying the new heat load $(\lambda)'$ from the back-pressure set, then

$$\left(\frac{1 - \dot{W}_C}{\eta_{CG}}\right)(1 - \eta_{CG}) = (\dot{\lambda})'(1 + \phi).$$

Hence
$$\Delta\dot{F}_b = \left(\frac{\dot{\lambda}_1}{\eta_{B_1}} + \frac{\dot{\lambda}_2}{\eta_{B_2}} + \frac{\dot{W}_S}{\eta_C}\right) - \left[\frac{(\dot{\lambda})'(1 + \phi)}{1 - \eta_{CG}}\right]\left(1 - \frac{\eta_{CG}}{\eta_C}\right),$$

and
$$\gamma_b = \left(\frac{\Delta\dot{F}}{\dot{\lambda}}\right)_b = \gamma_1 + \gamma_2 + \gamma_S - \frac{\delta(1 + \phi)}{(1 - \eta_{CG})}\left(1 - \frac{\eta_{CG}}{\eta_C}\right). \qquad (3.156)$$

Again a fuel energy savings ratio can be expressed in terms of the fuel energy originally required

$$
\begin{aligned}
(\text{FESR})'_b &= \frac{(\Delta \dot{F}/\dot{\lambda})_b}{\left[\dfrac{\dot{\lambda}_1}{\dot{\lambda}\eta_{B_1}} + \dfrac{\dot{\lambda}_2}{\dot{\lambda}\eta_{B_2}} + \dfrac{(1 + \dot{W}_S)}{\dot{\lambda}\eta_C}\right]} \\
&= \frac{\gamma_1 + \gamma_2 + \gamma_S - \dfrac{\delta(1 + \phi)}{(1 - \eta_{CG})}\left(1 - \dfrac{\eta_{CG}}{\eta_C}\right)}{\gamma_1 + \gamma_2 + \gamma_S + \dfrac{1}{\dot{\lambda}\eta_C}}.
\end{aligned}
\qquad (3.157)
$$

3.9.3 Supplying the Domestic Heating by a Pass-Out Turbine

Thirdly, we consider replacement of the existing mix of heating by heat rejected through the condensation of steam bled from an electrical power station turbine. Figure 3.33 shows the existing plant and the modified plant.

The extraction of steam to supply the heat load $(\dot{\lambda})'$ causes a reduction in the work output from the turbine, where Z is the so-called "Z-factor" given by

$$
Z = \frac{\text{Reduction in Work Output}}{\text{Useful Heat Rejected}}
$$

and discussed in section 2.6.2. The lost work has to be made up from a grid turbine supplying $\dot{W}_C$, but there is some compensation because electricity is not now being used for space heating as in the original heating "mix". The fuel energy originally supplied was

$$
\dot{F}_{REF} = \frac{\dot{\lambda}_1}{\eta_{B_1}} + \frac{\dot{\lambda}_2}{\eta_{B_2}} + \frac{(1 + \dot{W}_S)}{\eta_C}.
$$

If the CHP turbine plant essentially involves a modification of a grid turbine (through extraction of steam) and it now receives fuel energy $\dot{F}_{CG}$, then

$$
\eta_{CG}\dot{F}_{CG} = \eta_C\dot{F}_{CG} - Z(1 + \phi)(\dot{\lambda})' = 1 - \dot{W}_C,
$$

and the fuel energy supplied to the "make-up" grid turbine is $(\dot{W}_C/\eta_C)$. Hence the total fuel now supplied is

$$
\dot{F}_{CG} + \frac{\dot{W}_C}{\eta_C} = \frac{1}{\eta_C} + \frac{Z(1 + \phi)(\dot{\lambda})'}{\eta_C}.
\qquad (3.158)
$$

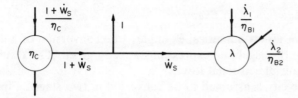

Reference plant

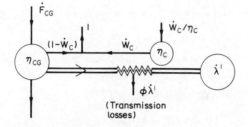

(Transmission losses)

Extraction turbine

FIG. 3.33. Fuel savings in district heating by extraction turbine.

The non-dimensional fuel saving (expressed in terms of the basic heat load) is then

$$\gamma_c = \left(\frac{\Delta \dot{F}}{\dot{\lambda}}\right)_c = \frac{\dot{\lambda}_1}{\dot{\lambda}_{B_1}} + \frac{\dot{\lambda}_2}{\dot{\lambda}\eta_{B_2}} + \frac{(1 + \dot{W}_S)}{\eta_C \dot{\lambda}} - \left[\frac{1}{\eta_C \dot{\lambda}} + \frac{Z(1 + \phi)(\dot{\lambda})'}{\eta_C \dot{\lambda}}\right],$$

(3.159)

and the fuel energy savings ratio, expressed in terms of the fuel energy originally required is

$$(\text{FESR})'_c = \frac{(\Delta \dot{F}/\dot{\lambda})_c}{\gamma_1 + \gamma_2 + \gamma_S + \left(\dfrac{1}{\dot{\lambda}\eta_C}\right)}$$

$$= \frac{\gamma_1 + \gamma_2 + \gamma_S - \dfrac{Z\delta(1 + \phi)}{\dot{\lambda}\eta_C}}{\gamma_1 + \gamma_2 + \gamma_S + \left(\dfrac{1}{\dot{\lambda}\eta_C}\right)}.$$

(3.160)

3.9.4 Calculation of an Example

Finally we give a numerical example—the conversion of a large city to CHP, using the algebraic expressions already given. The results are presented in non-dimensional form.

The large city is assumed to be converted in two stages. First electrical space heating $((\dot{W}_S)_{ON}$, on peak, and $(\dot{W}_S)_{OFF}$, off peak) and domestic heating $(\dot{\lambda}_1$, gas central heating, and $\dot{\lambda}_2$, other coal-fired heating) is converted to district heating by a central boiler plant. The non-dimensional fuel saving in this stage (a) of the conversion is

$$\gamma_a = \left(\frac{\Delta \dot{F}}{\dot{\lambda}}\right)_a = \gamma_1 + \gamma_2 + \gamma_S - (\gamma_B)'. \qquad (3.153b)$$

With
$$\dot{\lambda}_1 = 4.5 \text{ GJ p.a.,}$$
$$\dot{\lambda}_2 = 30 \text{ GJ p.a.,}$$
$$(\dot{W}_S)_{ON} = 8 \text{ GJ p.a.,}$$
$$(\dot{W}_S)_{OFF} = 3 \text{ GJ p.a.,}$$
$$\dot{\lambda} = 45.5 \text{ GJ p.a.,}$$
$$(\dot{\lambda})' = 50.05 \text{ GJ p.a.,}$$

and $\eta_{B_1} = 0.65$, $\eta_{B_2} = 0.5$, $(\eta_B)' = 0.85$, $\phi = 0.1$, $\eta_C = 0.3$; it follows that $\gamma_1 = 0.15$, $\gamma_2 = 1.32$, $\gamma_S = 0.81$, $(\gamma_B)' = 1.42$, $\delta = 1.1$ and $\gamma_a = 0.86$.

Thus 86% of the energy required in the dwelling is saved by the conversion to "heat only" boilers. The critical point here is the low value of efficiency assumed for η_{B_2}—coal fired heating—in the original system.

The second stage (d) involves conversion from "heat only" boilers to pass-out steam heating. The non-dimensional fuel saving is

$$\gamma_d = \left(\frac{\Delta \dot{F}}{\dot{\lambda}}\right)_d = (\gamma_B)' - \frac{Z\delta(1 + \phi)}{\eta_C}, \qquad (3.161)$$

assuming no provision for stand-by boiler operation.

With
$$Z = 0.14, \; \eta_C = 0.3, \; \phi = 0.1, \; \delta = 1.1,$$

$$\gamma_d = 1.42 - \frac{0.14 \cdot 1.1 \cdot 1.1}{0.3} = 0.855.$$

The total (non-dimensional) fuel savings in the two stages are

$$\gamma_c = \gamma_a + \gamma_d$$

$$= (\gamma_1 + \gamma_2 + \gamma_S - (\gamma_B)') + \left((\gamma_B)' - \frac{Z\delta(1 + \phi)}{\eta_C}\right)$$

$$= 0.86 + 0.855$$

$$= 1.715.$$

Figure 3.34 shows the various components of energy savings in these two stages. The left-hand side of the figure shows the heat savings in the build-up of the heat load, taken over by heat only boilers. The right-hand side of the figure shows the further substantial energy savings that occur when that heat load is accepted by the pass-out turbine, from the heat-only boilers.

It should be appreciated that the fuel savings are indeed greater than the heat requirement (λ) of the original "average" house. This is because the efficiency of the initial heating scheme was low, and the initial energy input to the original system was (non-dimensionally)

$$(\dot{F}_{REF}/\lambda) = 0.15 + 1.32 + 0.81 = 2.28.$$

Further the final heating system involves use of heat originally wasted by the power station, so it is not surprising that γ exceeds unity. There is no contravention of the Second Law of Thermodynamics involved in such a result.

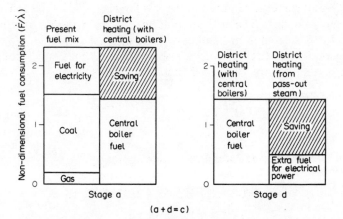

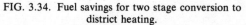

FIG. 3.34. Fuel savings for two stage conversion to district heating.

References

1. Porter, R. W. and Mastanaiah, K. Thermal-Economic Analysis of Heat-Matched Industrial Cogeneration Systems. *Energy*, **7**, 2, 171–187, 1982.
2. Bammert, K. Twenty-Five Years of Operating Experience with the Coal-Fired Closed-Cycle Gas Turbine Cogeneration Plant at Coburg. *Trans. ASME Journal of Engineering for Power*, **105**, 806–815, 1983.
3. Campbell, J., Lee, J. C., and Wright, D. E. Coal-Fired Closed-Cycle Gas Turbine Cogeneration Systems. ASME Paper 80-GT-156, 1980.
4. Kolbusz, P. The Use of Heat Pumping in District Heating Systems. Electricity Research Council Report no. ECRC/M700, 1974.
5. Casci, C. and Gaia, M. Heat Pump Enhanced Gas Turbine Cogeneration. *Energy*, **9**, 7, 555–564, 1984.
6. Horlock, J. H. Combined Heat and Power Supply Using Carnot Engines. *Liber Amicorum*, Andrew L. Jaumotte. Note Technique 50 de l'Institute de Mecanique Appliquee, 1983.
7. Haywood, R. W. Private communication, 1986.
8. Oliker, I. Steam Turbines for Cogeneration Power Plants. *Trans. ASME Journal of Engineering for Power*, **102**, 482–485, 1980.
9. Timmermans, A. R. J. Combined Cycles and their Possibilities. Lecture Series, Combined Cycles for Power Generation. Von Karman Institute for Fluid Dynamics, Rhode Saint Genese, Belgium, 1978.
10. Lowder, J. R. A. Improved Matching of Gas Turbines for CHP Using Recuperators. GEC Power Engineering Report, 1977. (See also Ref. 9, Ch. 4.)
11. Horlock, J. H. The Efficiency and Performance of Rankine-type Steam Cycles. *Proc. Instn. Mech. Engrs.*, **201**, A3, 1987.
12. Salisbury, J. K. The Steam-Turbine Regenerative Cycle—An Analytical Approach. *Trans. ASME*, **64**, 231, 1942.
13. Haywood, R. W. *Analysis of Engineering Cycles* (3rd Edition). Pergamon Press, Oxford, 1980.
14. Horlock, J. H. Analysis of the Energy and Money Savings Associated with CHP Plant. District Heating Combined with Electricity Generation in the United Kingdom (Appendix 12). Department of Energy. HMSO, London, 1977.

CHAPTER 4

Economic Assessment of CHP Schemes

4.1 Introduction

In Chapter 3, details were given of the thermodynamic assessment of various CHP schemes. The work and useful heat output of a CHP plant, for unit energy input through the fuel supply, may be estimated with fair accuracy. Given the heat and work loads to be met (and their ratio λ_D), the energy supply required (and the rate of supply of fuel) may then be assessed. Further, in comparison with "basic" separate plants meeting the same loads (usually an existing "mix" of electrical power taken from the grid and a heat load met from a variety of sources (electric space heating; coal heating; coal, gas or oil boilers)), the energy savings from a CHP scheme may be obtained relatively simply and with fair accuracy (as described in section 3.9 for a district heating scheme). These savings are usually substantial, for the original, often inefficient, heat sources are not used. Heat hitherto wasted (by rejection to the environment from the power plant supplying the grid) is utilised instead.

If the *thermodynamic* advantages of CHP plant are so clear, why is CHP not universally adopted? The answer lies in the study of the *economics* of heat and power, and particularly whether the capital expenditure required to install a new plant is justified. Such capital expenditure will involve not only a new and different type of power plant, but also the costs of major and minor heating networks. In big district heating schemes these include the installation of large pipes to carry hot water from a power station to a town where the district heating load exists; the subsidiary distribution of that hot water; and the central heating pipework in the individual houses.

This book is a volume for engineers, concerned primarily with the thermodynamics of CHP. But the economics of a proposal must be studied in order to explain under what conditions CHP plant may be installed in future. Indeed many such plants exist, and the economics of their operation may also be considered.

In this chapter (and in Appendix B) we give a brief account of various methods of assessing the economics of CHP plant, together with some examples. Some answers can be obtained quickly (e.g. simple assessment of pay-back period based on initial capital cost). Methods involving con-

111

sideration of the time value of money, and the concept of discounted cash flow, are more complex. If capital input and savings vary from year to year computers are required to give accurate answers. But even for some such cases, simple analyses may be derived (see Appendix B) and these are used in this chapter. Each method of analysis involves the *comparison* of a new scheme with another (usually existing) scheme, either explicitly or implicitly.

First, in section 4.2, we describe two methods of pricing the output from a CHP plant:

(i) by pricing the electricity produced, taking financial credit for the elimination of the "auxiliary" supply of heat, a method often used for industrial CHP schemes;

(ii) by pricing the useful heat produced, assuming values for electricity bought or sold, a method more frequently used for district heating CHP applications.

In both cases explicit comparisons of the calculated prices for a *new* CHP plant are made with *existing* prices (of electricity or heat respectively), but "demand side" economics (the effect of price changes on demand) is neglected. Capital charges are included in these pricing comparisons.

Next, in section 4.3, after a brief description of "traditional" assessments of short pay-back periods, the techniques of discounted cash flow are used, usually (but not always) to describe the interaction between capital expenditure and resulting fuel (and money) savings achieved by *changing* from an existing system to a new CHP plant, a more implicit means of comparison. The savings from improved utilisation of energy less the extra capital and other expenditure required may be discounted back to the start of the operation, and a net present value (NPV) (or a levelised annual value of the cash flow) determined. Alternatively, but from the same type of analysis, a more accurate assessment may be made of the discounted pay-back period (DPBP), of the break-even capital cost (BECC), or the investors' rate of return (ROI). All these are ways of expressing a similar answer on the economics of the plant, and since different authors have used different criteria we describe them briefly here, quoting results from each approach.* A list of these authors, with the criteria used, is given in Table 4.1; the type of CHP plant assessed (district heating (DH) or industrial (IND)) is also indicated, and whether the direct comparisons are made, or a "change" type of analysis is used.

---

*But note that the economic theory of cost benefit analysis points to some significant differences of principle between the applications of (NPV), (BECC) and (ROI), if "maximisation of utility" is the objective (reflecting the preference of people for money and goods *now*, rather than in the future).

TABLE 4.1

Author	Criterion for assessment	Industrial (CHP/IND) or district heating (CHP/DH)	Type of assessment (direct comparison (DC) or "change" (C))
Marshall[1] (small city)	Net present value (NPV) or levelised annual value of cash flow (Ā)	CHP/DH	DC
Marshall[1] (medium city)	Net present value (NPV) or levelised annual value of cash flow (Ā) .	CHP/DH	C
Marshall[1] (large city)	Net present value (NPV) or levelised annual value of cash flow (Ā)	CHP/DH	C
Williams[2]	Electricity price	CHP/IND	DC
Belding[3]	Electricity price	CHP/IND	DC
Kehlhofer[4]	Electricity price	CHP/IND	DC
Kehlhofer[4]	Heat price	CHP/DH	DC
Lilley[5]	Simple pay-back period (PBP)	CHP/IND	C
Jody et al.[6]	Venture worth (NPV), break-even capital cost (BECC)	CHP/IND	C
Daudet and Trimble[7]	DCF or investor's rate of return (ROI)	CHP/IND	C
Porter and Mastanaiah[8]	DCF or investor's rate of return (ROI)	CHP/IND	C
Flint and El-Masri[9]	DCF or investor's rate of return (ROI)	CHP/IND	C

The author has been associated primarily with the use of NPV (in an extensive study of CHP by the Marshall Committee[1]); in the latter part of this chapter particular emphasis is placed on this approach. Following the examples of fuel saving, given at the end of Chapter 3, simple analyses are given for the determination of NPV in three cases (use of heat only boilers, back pressure turbines or pass-out turbines to replace existing conventional plant). The results of the comprehensive calculations carried out by the Marshall group are described in section 4.4. However, the results of other authors, using other criteria (mainly rate of return (ROI) and break-even capital cost (BECC)), are also discussed in this section. Finally, the effect of unmatched operation on the economics of CHP plant is discussed in section 4.5.

4.2 Direct Pricing of Electricity and Heat

Pricing of *electricity* from any power station is a complex business; as several authors have made economic studies of CHP plant in this way, we describe their methods and results here (section 4.2.1). We also describe an alternative approach, that of pricing the *heat output* from a CHP scheme (section 4.2.2).

4.2.1 *Pricing of Electricity*

Williams[2] and Belding[3] describe studies of the economics of CHP (cogeneration) which essentially involve a direct comparison between the pricing of electricity from CHP plant with that supplied from existing central power stations. Kehlhofer[4] follows a similar approach, which has validity mainly for independent cogeneration of electricity and heat for industrial processes (CHP/IND). These methods are based on relating electricity prices to both capital related costs and the recurrent costs of production (fuel and maintenance of plant):

$$P_E = \beta C + M + (OM), \tag{4.1}$$

where P_E is the annual cost of the electricity produced (£ or $ per annum);

C is the capital cost of plant (e.g. £ or $);

β (i, N) is the capital charge factor which is related to the discount rate (i) on capital and the life of the plant (N years);

M is the annual cost of fuel supplied (e.g. £ or $ p.a.);

$\underline{(OM)}$ is the annual cost of operation and maintenance (e.g. £ or $ p.a.).

A full description of the derivation of the capital charge factor is given in Appendix B.

The straightforward way of assessing a "matched" CHP scheme (subscript CG) is then to compare its electricity production cost

$$(P_E)_{CG} = \beta C_{CG} + M_{CG} + (OM)_{CG} \tag{4.2}$$

directly with that of a conventional or "basic" plant (subscript C)

$$(P_E)_C = \beta C_C + M_C + (OM)_C. \tag{4.3}$$

However, the CHP plant is doing more than meeting an electrical power load. It is also supplying heat to process or district heating, which would otherwise have been supplied by an "auxiliary" heat source (subscript B) at an annual cost of

$$\beta C_B + M_B + (OM)_B.$$

That cost has been saved, so that if the electrical power output is $\dot{W}$ and the plant operates for H hours per annum, the net "unitised" production cost of electricity (say £ per kilowatt hour) for the cogeneration plant should be

$$(Y_E)_{CG} = \frac{(P_E)_{CG}}{\dot{W}H} = \frac{\beta(C_{CG} - C_B)}{\dot{W}H} + \frac{(M_{CG} - M_B)}{\dot{W}H} + \frac{[(OM)_{CG} - (OM)_B]}{\dot{W}H}.$$

(4.4)

The costs M may be written as the product of the unit price of fuel ($\mathscr{S}$, £/kWh),* the rate of supply of energy in the fuel ($\dot{F}$, kW), and the utilisation factor H (hours p.a.),

$$\frac{M_{CG} - M_B}{\dot{W}H} = \frac{(\mathscr{S}_{CG}\dot{F}_{CG} - \mathscr{S}_B\dot{F}_B)H}{\dot{W}H} = \frac{(\mathscr{S}_{CG}\dot{F}_{CG} - \mathscr{S}_B\dot{F}_B)}{\dot{W}}. \quad (4.5)$$

If the cost of fuel ($\mathscr{S}_{CG}$) is low, as in a plant incinerating waste, then clearly the production cost $(Y_E)_{CG}$ will come down; however, even in such a plant fuel costs are finite, because of handling and transportation.

More generally, the quantity $\dot{F}_B$ is equal to $(\dot{Q}_U/\eta_B)$ where $\dot{Q}_U$ is the heat load (expressed in terms of energy rates, MW or kW) and η_B is the "boiler" efficiency of the original displaced heat sources. The quantity $\left(\dfrac{\dot{W}}{\dot{F}_{CG} - (\dot{Q}_U/\eta_B)}\right)$ is the artificial efficiency defined in section 2.3.3 (equation (2.5)) or, as Kehlhofer calls it, the efficiency of power generation (η_a). Williams calls $[\dot{F}_{CG} - (\dot{Q}_U/\eta_B)]/\dot{W}$ the effective heat rate of the plant, and Porter and Mastanaiah call it the incremental heat rate $(IHR)_{CG}$ (if $(\eta_B)_{CG} = \eta_B$). Then if the costs of fuel to the new plant and the displaced heat sources are the same, $\mathscr{S}_{CG} = \mathscr{S}_B = \mathscr{S}$,

$$(Y_E)_{CG} = \frac{\beta(C_{CG} - C_B)}{\dot{W}H} + \frac{\mathscr{S}}{\eta_a} + \frac{[(OM)_{CG} - (OM)_B]}{\dot{W}H} \quad (4.6a)$$

or

$$(Y_E)_{CG} = \frac{\beta(C_{CG} - C_B)}{\dot{W}H} + \mathscr{S}(IHR)_{CG} + \frac{[(OM)_{CG} - (OM)_B]}{\dot{W}H}.$$

(4.6b)

This unitised production cost may be compared with that for the basic plant,

$$(Y_E)_C = \frac{(P_E)_C}{\dot{W}H}. \quad (4.7)$$

---

*Note that $\mathscr{S}$(£/kWh) is equal to S (£/kg) divided by the calorific value (CV) (kWh/kg), i.e. $\mathscr{S} = S/(CV)$.

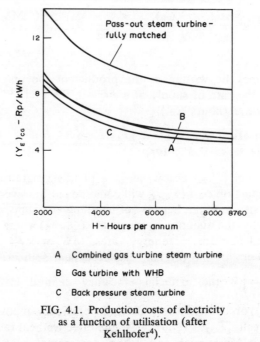

A   Combined gas turbine steam turbine

B   Gas turbine with WHB

C   Back pressure steam turbine

FIG. 4.1. Production costs of electricity
as a function of utilisation (after
Kehlhofer[4]).

Figure 4.1 shows Kehlhofer's calculation of production cost $(Y_E)_{CG}$ as a function of the utilisation period (H) for an industrial CHP plant, using a pass-out steam turbine designed to match the demands of 45 MW (electricity) and provision of 25 kg/s of process steam at 3.5 bar/190°C. Also shown are values of $(Y_E)_{CG}$ for three other CHP plants (a combined gas/steam turbine plant (A), a gas turbine with a waste heat boiler (B), and a back pressure steam turbine (C)) which are designed to match the process heat load, but do not match the electrical demand of 45 MW (so that extra electrical power will have to be bought in). Clearly all the production costs come down substantially with increased utilisation. Table 4.2 gives the break-down in electricity production costs for the four plants.

Kehlhofer also gives the average electricity *price* $\bar{Y}_E$ (presumably equating sale price to production cost, without allowance for tax or profit) if power has to be bought in, at a price $(Y_E)_{BI}$.

If the electrical demand is $\dot{W}_D$ (for the utilisation period H) and that supplied is $\dot{W}_{CG}$ (also for H hours) then this price is

$$\bar{Y}_E = \frac{\dot{W}_{CG}H(Y_E)_{CG} + (\dot{W}_D - \dot{W}_{CG})H(Y_E)_{BI}}{\dot{W}_D H}$$

$$= \frac{\dot{W}_{CG}(Y_E)_{CG}}{\dot{W}_D} + \left[1 - \left(\frac{\dot{W}_{CG}}{\dot{W}_D}\right)\right](Y_E)_{BI} \qquad (4.8)$$

TABLE 4.2. *Economic Comparison of CHP/IND Plants* (after Kehlhofer[4])

		Extraction/ condensing turbine	Back- pressure turbine	Gas turbine with heat recovery boiler	Combined plant
Net power MW output	MW	45.0	15.2	25.8	39.1
EUF		0.43	0.81	0.70	0.75
Additional capital cost over steam boiler	S.fr. (millions)	38.0	10.4	18.0	27.2
Production cost of electricity	Rp./kWh	8.59	4.90	5.44	5.32
Capital-dependent costs*	Rp./kWh	1.8	1.46	1.49	1.48
Operating costs*	Rp./kWh	0.69	0.21	0.22	0.36
Fuel costs*	Rp./kWh	6.1	3.23	3.73	3.48

Boundary conditions:
—Process heat flow   = 25 kg/s (90 t/h)
—Process steam conditions   = 3.5 bar/190°C
—Power demand   = 45 MW
—Equivalent utilisation period (H)   = 7000 h/a
—Annuity factor (β)   = 0.14 (10 years, 8% interest)
—Fuel price ($\mathscr{S}$)   = 7.3 S.fr./GJ (natural gas)

*Difference from simple steam boiler with electricity purchased from network
1 S.fr. (Swiss francs) ≡ 100 R

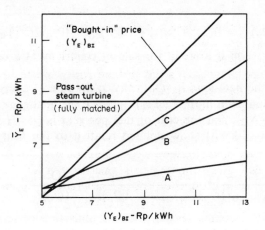

A   Combined gas turbine steam turbine

B   Gas turbine with WHB

C   Back pressure steam turbine

FIG. 4.2. Average cost of electricity as
a function of "bought-in" price (after
Kehlhofer[4]).

and increases linearly with $(Y_E)_{BI}$, as Fig. 4.2 shows (except for the pass-out turbine for which $\dot{W}_D = \dot{W}_{CG}$). If the demand varies over the utilisation period then a weighted price can be determined.

Similar calculations by Williams[2] also emphasise the extra cost of bought-in electricity, which has to be purchased from the grid if the industrial cogeneration fails or if electricity demand exceeds the cogeneration supply. Williams includes standby charges for the cogeneration plants, on the basis of 10% of on-site need and a peak requirement equal to half the peak industrial demand. These assumptions set the cost of standby electricity per kWh at about double that which an industry would pay for full grid service, and push CHP prices above normal grid prices (but less than that from nuclear plant in the USA).

4.2.2 Pricing of Heat

An alternative assessment of CHP plant is relevant for district heating; here, for the independent producer, the approach is to price the *heat*, assuming the electricity to be a by-product which can be sold to the grid. The sale (or buy-back) price, (like the stand-by (or bought-in) cost in the industrial CHP examples of section 4.2.1) is then a matter for negotiation with the electricity utility.

The cost of heat (£ p.a.) produced from the cogeneration plant is now given by

$$(P_H)_{CG} = \beta C_{CG} + M_{CG} + (OM)_{CG} - \dot{W}_{CG}H(Y_E)_{BB}, \qquad (4.9)$$

where the last term is a credit for sale of electricity at a buy-back price $(Y_E)_{BB}$.

As before, the heat load is $(\dot{Q}_U)_D$ (kW); the rate of energy supplied to the cogeneration plant in the fuel is $\dot{F}_{CG}$ (kW); the heat generated per year is $(\dot{Q}_U)_{CG}H$ (kWh); and the cost of fuel per year is $(\dot{F}_{CG}H\mathscr{S}_{CG})$. The net (unitised) cost (£/kWh) (now defined relative to the *heat* supplied per year, is

$$(Y_H)_{CG} = \frac{(P_H)_{CG}}{(\dot{Q}_U)_{CG}H} = \frac{\beta C_{CG}}{(\dot{Q}_U)_{CG}H} + \frac{\dot{F}_{CG}\mathscr{S}_{CG}}{(\dot{Q}_U)_{CG}} + \frac{(OM)_{CG}}{(\dot{Q}_U)_{CG}H} - \frac{\dot{W}_{CG}(Y_E)_{BB}}{(\dot{Q}_U)_{CG}},$$

$$(4.10)$$

compared with the cost of heat supplied by "heat only" boilers

$$(Y_H)_B = \frac{(P_H)_B}{(\dot{Q}_U)_{CG}H} = \frac{\beta C_B}{(\dot{Q}_U)_{CG}H} + \frac{F_B\mathscr{S}_B}{(\dot{Q}_U)_{CG}H} + \frac{(OM)_B}{(\dot{Q}_U)_{CG}H}. \quad (4.11)$$

A similar approach is followed by Marchant, Proost and Wilberg,[10] but it involves more complexity; time-variable sale prices of electricity are

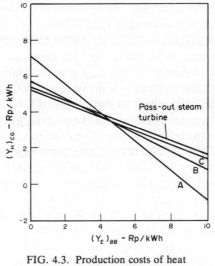

FIG. 4.3. Production costs of heat
as a function of electricity "buy-
back" price (after Kehlhofer[4]).
Combined gas turbine/steam
turbine (A), gas turbine with WHB (B),
back pressure steam turbine (C).

included together with standby costs, in case of plant failure. Heat prices
clearly cannot be based solely on thermodynamic considerations alone, for
other factors are vitally important. Not only do capital and maintenance
costs have a major influence but also the sale (or buy-back) price of
electricity.

Kehlhofer[4] gives production costs for a heat load of 60 MW met by the
CHP plants listed earlier. Figure 4.3 shows the cost of heat plotted against
sale price of electricity, $(Y_E)_{BB}$. For a high buy-back price the combined
plant is most attractive, but with a low value of $(Y_E)_{BB}$ the back pressure
turbine is the most economic.

4.2.3   A Note on the Pricing of Electricity or Heat Supplied by a Utility

If the vendor of heat or electricity is an electrical utility (rather than an
independent producer) the approaches outlined in sections 4.2.1 and 4.2.2
have to be reconsidered. That adopted in section 4.2.1 for pricing electricity remains largely valid, but the *cost* of "bought-in" electricity (see, for
example, equation (4.8)) and/or of standby provision become an internal
question for the utility. Further, while the credit for "displacing" the heat
load (see equation (4.5)) may be valid for a demand met within the utility

itself, more usually the credit will appear through a sale to an external consumer in industrial cogeneration (CHP/IND).

More important is the question of pricing the district heating met through cogeneration (CHP/DH) by an electrical utility (local or national). The approach of section 4.2.2 has to be modified, because taking a credit for the sale of electricity is no longer strictly valid. The method adopted by the Marshall Group[1] in their study of CHP/DH was to assume that the overall electrical demand (the sum of that supplied by the new CHP or cogeneration station and that "externally" connected to the grid) had to be provided by the utility. It was then logical to price the heat supplied for comparison with other methods (e.g. by gas for central heating, or by a mix of fuels (say) in a large city), making sure there was no alteration in the price of electricity, for the overall utility system. Alternatively, in a "change" type of analysis, the extra cost or the net savings from producing heat through cogeneration, in place of an existing heating system, could be estimated.

This approach is described later in this chapter, in section 4.4.

4.3   Methods of Analysis Using Discounted Cash Flow (DCF) Techniques (Rate of Return, Pay-back Period, Net Present Value, Break-even Capital Cost)

The concept of time value of money is implicit in the introduction of capital charges (βC) into the pricing of electricity and heat, as described in section 4.2. More explicit use of the concept is involved in a variety of discounted cash flow (DCF) techniques, particularly the net present value (or venture worth) used extensively in the subsequent sections of this chapter, and described in detail in Appendix B. However, before studying these DCF techniques, it is useful to describe two "traditional" criteria of profitability often used for quick assessment of CHP schemes. They make no allowance for the time value of money, but give a simple introduction to economic assessment. Thus

(i) simple rate of return
(% rate of return) $= \dfrac{\text{(Annual return)}}{\text{(Invested capital)}} \times 100;$

(ii) simple pay-back period $= \dfrac{\text{(Invested capital)}}{\text{(Annual return)}}.$

Lilley[5] suggests that in considering short pay-back periods, which industry usually requires, these concepts are often sufficient. Thus for a site with "demand" heat load $(\dot{Q}_U)_D H$ (kWh), met conventionally by a boiler of efficiency η_B using fuel at price $\mathscr{S}_B$, and a "demand" electricity load

$\dot{W}_D H$ (kWh), met by power bought from the grid at price $(Y_E)_{BI}$, the annual energy costs are

$$M_C = \left[(Y_E)_{BI}\dot{W}_D + \frac{\mathscr{S}_B(\dot{Q}_U)_D}{\eta_B}\right]H. \tag{4.12}$$

If the conventional scheme is replaced by a cogeneration (CG) scheme, of efficiency η_{CG}, producing electrical power $\dot{W}_{CG}$ and heat $(\dot{Q}_U)_{CG}$, and using fuel priced at $\mathscr{S}_{CG}$, then energy costs (£ or $ p.a.) are

$$M_{CG} = \left[\frac{\mathscr{S}_{CG}\dot{W}_{CG}}{\eta_{CG}} + (\dot{W}_D - \dot{W}_{CG})(Y_E)_{BI}\right]H,$$
$$\text{if} \quad \dot{W}_{CG} < \dot{W}_D, \; (\dot{Q}_U)_{CG} = (\dot{Q}_U)_D; \tag{4.13a}$$

$$M_{CG} = \left[\frac{\mathscr{S}_{CG}\dot{W}_{CG}}{\eta_{CG}} + [(\dot{Q}_U)_D - (\dot{Q}_U)_{CG}]\,\mathscr{S}_B\right]H,$$
$$\text{if} \quad \dot{W} = \dot{W}_D, \; (\dot{Q}_U)_{CG} < (\dot{Q}_U)_D; \tag{4.13b}$$

$$M_{CG} = \left[\frac{\mathscr{S}_{CG}\dot{W}_{CG}}{\eta_{CG}} - (\dot{W}_{CG} - \dot{W}_D)(Y_E)_{BB}\right]H,$$
$$\text{if} \quad \dot{W}_{CG} > \dot{W}_D, \; (\dot{Q}_U)_{CG} = (\dot{Q}_U)_D; \tag{4.13c}$$

where $(Y_E)_{BB}$ is the sale price of electricity (the buy-back price) and $(Y_E)_{BI}$ the purchase (buy-in) price. (Note that if $(\dot{Q}_U)_{CG} > (\dot{Q}_U)_D$, then $[(\dot{Q}_U)_{CG} - (\dot{Q}_U)_D]$ is heat that will be wasted, not sold, so a fourth level of energy costs does not arise). The money savings (resulting from the fuel savings as described in Chapter 3) are therefore

$$\Delta M = M_C - M_{CG}$$
$$= [(Y_E)_{BI}\dot{W}_D + (\mathscr{S}_B\dot{Q}_U/\eta_B)]H - M_{CG}, \tag{4.14}$$

but these may be reduced by $\Delta(OM)$, the *increased* cost of operation and maintenance.

The percentage rate of return is then $100\left(\dfrac{\Delta M}{\Delta C}\right)$ p.a., where ΔC is the additional invested capital for the new CHP scheme, and the pay-back period is $(\Delta C/\Delta M)$.

These simple assessments can also take account of part-load operation (Lilley[5]) and we discuss this aspect later, in section 4.5.

Discounted cash flow techniques modify the "static" values of money used in these simple assessments. For example the pay-back period has to be modified to allow for capitalised costs. This and other assessments of the economic viability of CHP schemes are given in the next sections and are based on the discussion of the time value of money given in Appendix B.

It should be emphasised again that in these economic studies, it must be clear whether either (i) the "absolute" cost of the new project is to be determined, and prices *compared directly* with those for an existing conventional scheme,

or (ii) the *money saving* (probably in the form of the net present value), related to a *change* from an existing scheme is to be assessed.

In the direct pricing method described in section 4.1, the first approach was adopted, i.e. the prices of supply in new and existing schemes were compared, each taking account of the total capital charges. In the simple estimates of pay-back period and rate of return just described, the second approach was followed, i.e. the savings associated with a *change* from an existing system (with an existing mix of plant and fuel supplies) to a new CHP scheme were considered, and it is this approach which is usually (but not always) followed in DCF analysis. The use of this second ("change") method involves the assumption that a large electricity supply grid is available (surplus electricity from the CHP plant can be sold to the grid, or electricity can be bought from it if local demand exceeds the CHP generated electricity); or that additional boilers may be introduced (if the CHP plant cannot meet the local heat demand). As in the examples given of direct pricing of electricity or heat, the value of these supplementary supplies (electricity or heat) are important factors in DCF analyses.

4.3.1 Net Present Value (NPV)

The net present value derived from a "change" type of analysis—from an existing conventional scheme to a new CHP scheme—balances extra capital cost (ΔC) against the rate of savings which arise (ΔM), mainly from the reduced use of fuel. We list here a series of examples of increasing complexity, the full details of which are given in Appendix B; corresponding equation numbers from the appendix are given. The use of "Δ" emphasises that the "change" type of analysis is being adopted.

4.3.1.1 Simple Analysis

(i) In the first case, "extra" capital ΔC invested at time zero leads to immediate annual money saving ΔM in the first year and this money is invested at the end of that year. Subsequent years lead to identical savings and investment. These savings may be discounted back to time zero and lead to a net present value of

$$(\text{NPV})_{(i)} = \frac{\Delta M}{(_{\text{N}}f_{\text{AP}})} - \Delta C, \qquad (4.15)$$
$$(\text{B.1})$$

where $(_N f_{AP})$ is the annuity present worth factor, for an interest rate i over a period of N years, the life of the project,

$$(_N f_{AP}) = \frac{i(1 + i)^N}{[(1 + i)^N - 1]}.$$

(4.16)
(B.2)

(ii) If the capital is invested l years ahead of the resultant money savings then the net present value at time zero is

$$(NPV)_{(ii)} = \frac{\Delta M}{(_N f_{AP})} - \Delta C(1 + i)^l.$$

(4.17)
(B.3)

(iii) If the capital is spent linearly over L years, ahead of year zero (i.e. annual spend $(\Delta C/L)$) then the net present value at time zero is

$$(NPV)_{(iii)} = \frac{\Delta M}{(_N f_{AP})} - \frac{\Delta C}{L(_L f_{AF})},$$

(4.18)
(B.4)

where $(_L f_{AF})$ is the *future* worth factor over L years,

$$(_L f_{AF}) = \frac{i}{[(1 + i)^L - 1]}.$$

(4.19)
(B.5)

(iv) Finally, if capital is invested as in case (iii), but money savings also build up (from time zero) over a period of L years (increasing by $(\Delta M/L)$ per year) then the net present value at time zero is

$$(NPV)_{(iv)} = \frac{1}{L(_L f_{AP})}\left[\frac{\Delta M}{(_N f_{AP})} - \Delta C(1 + i)^l\right].$$

(4.20)
(B.6)

If the lead time on capital investment (l) were reduced to zero, and capital spend and money savings were in phase, then

$$(NPV)_{(iv)} = \frac{1}{L(_L f_{AP})}\left[\frac{\Delta M}{(_N f_{AP})} - \Delta C\right]$$

$$= \frac{(NPV)_{(i)}}{L(_L f_{AP})}.$$

(4.21)
(B.7)

Examples of the use of these simple analyses for NPV are given in section 4.4 (together with a summary of parallel comprehensive computer calculations of NPV) for some district heating schemes. These analyses were originally undertaken by the author for the Marshall Committee on combined heat and power, and appeared in an early report by its working party on district heating.[11]

It is of interest to relate the (NPV) obtained from a "change" analysis (leading say to equation (4.15)) to direct comparisons. If the conventional system (subscript C) has a net present cost $(NPV)_C$ then

$$(NPV)_C = \frac{M_C}{{}_Nf_{AP}} + C_C. \qquad (4.22)$$

But the cogeneration scheme (CG) has a net present cost

$$(NPV)_{CG} = \frac{M_{CG}}{{}_Nf_{AP}} + C_{CG}. \qquad (4.23)$$

The change from C to CG yields a positive net present value (saving)

$$\Delta(NPV) = (NPV)_C - (NPV)_{CG}$$

$$= \frac{(M_C - M_{CG})}{{}_Nf_{AP}} - (C_{CG} - C_C)$$

$$= \frac{\Delta M}{{}_Nf_{AP}} - \Delta C,$$

which is equation (4.15) with
$\Delta M = M_C - M_{CG}$, the annual *savings*,
and $\Delta C = C_{CG} - C_C$, the increased capital *cost*,
both positive quantities.

We may also note that the equations of the form of (4.22) and (4.23) may be written

$$_Nf_{AP}(NPV) = M + C(_Nf_{AP}),$$

which may be compared with equations (4.2) and (4.3). $_Nf_{AP} = \beta$ is the capital cost factor used in those equations to give the production costs. (These are also called levelised costs, as we shall discuss in section 4.3.1.3.)

4.3.1.2 Complex Estimates of NPV (and Associated Other Criteria)

Jody et al.[6] refer to NPV as "venture worth"; they present the results of their calculations in the form of break-even capital cost. They also calculate on a "change" basis, estimating the annual money savings (ΔM) from installation of a cogenerating system replacing a conventional plant, and balance this against the extra capital cost (ΔC). Like Williams,[2] in his estimate of electricity prices, Jody et al.[6] use complex rather than simple expressions for discount rate, allowing for tax (t), tax credit for depreciation (t_{CD}) and for investment (t_{CI}), and for minimum required return on high-risk ventures $(i)'$.

Their derivation of venture worth then becomes

$$(\text{NPV})_{(v)} = \left\{ \sum_{k=1}^{N} \frac{\Delta M(1 - t)}{(1 + i)^k} - \Delta C \right\}$$
$$+ \sum_{k=1}^{N} \left\{ \frac{t_{CD}}{(1 + i)^k} + t_{CI} - \frac{[(1 + i)^k - 1][(i)' - i]}{i(1 + i)^k} \right\} \Delta C.$$

(4.24)

This may be compared with the simple expression for $(\text{NPV})_{(i)}$ given in equation (4.15). The first term becomes identical with $(\text{NPV})_{(i)}$ when $t = 0$. Allowance may also be made for working capital and for salvage value of plant, but these effects are small. Allowance for inflation may be made in the choice of (effective) interest or discount rate. An effective rate of return used by Williams[2] in pricing electricity is approximately that with zero inflation less the percentage annual inflation.

For a *selected* discount rate (i) $(\text{NPV})_{(v)}$ may be set equal to zero to obtain the *break-even capital cost* (£ per kilowatt) as

$$(\text{BECC}) = \frac{\Delta C}{\dot{W}} = \frac{(1 - t) \sum\limits_{k=1}^{N} \dfrac{(\Delta M/\dot{W})}{(1 + i)^k}}{1 - \left\{ \sum\limits_{k=1}^{N} \dfrac{t_{CD}}{(1 + i)^k} - t_{CI} + \dfrac{[(1 + i)^k - 1][(i)' - i]}{i(1 + i)^k} \right\}}$$

(4.25)

where $\dot{W}$ is the electrical power output from the plant. Some results from their calculations are presented later in section 4.4.

Other authors[7,8] have followed similar approaches. Daudet and Trimble[7] determine net present value, but use it to calculate the DCF *rate of return*—the interest rate which gives a net present value of zero, for a *prescribed* capital investment.

Porter and Mastanaiah[8] also present results on the basis of the DCF or investors rate of return (in their phrasing, "the discounted after-tax rate of return on investment (ROI)—the interest rate which nulls the net discounted cash flow on either a present worth or net levelised annual basis"). They use the alternative concept of "levelised" cash flows, a description of which now follows.

4.3.2 Levelised Cash Flows

Levelised cash flows take account of variations in cash flow from year to year, essentially evening them out.

Consider a cash flow A_k in a given year (k). This will be made up of the money savings on fuel (resulting from a change from a conventional plant to a CHP scheme) less the capital and other costs (e.g. tax) in that year. Generally A_k will vary with time. The levelised cash flow ($\bar{A}$) is defined as

$$\bar{A} = {}_N f_{AP} \sum_{k=1}^{N} \left[\frac{A_k}{(1 + i)^k} \right], \tag{4.26}$$
$$\text{(B.14)}$$

and is the hypothetical (constant) cash flow which gives the same net present value as the real (variable) cash flow, for the same interest (i) and the same plant life (N years) (see Appendix B). The bar superscript thus indicates that the cash flows have been "levelised"—brought back to present worth and then evened out over the life of the plant, N years. (This levelised technique may also be used in determining prices of heat and electricity.)

The net present value is related to the levelised cash flow,

$$(\text{NPV}) = \frac{\bar{A}}{{}_N f_{AP}}. \tag{4.27}$$
$$\text{(B.13)}$$

The discounted cash flow rate of return (ROI), the break-even capital costs (BECC) or the pay-back period (PBP) may then be determined by putting either net present value (NPV) or the levelised cash flow (A) equal to zero.

Porter and Mastanaiah[8] use the levelised cash flow, putting it equal to zero in order to obtain the investors rate of return (ROI). In their calculations of the cash flow in any year, they emphasise that this must take account of both the cost of producing electricity in the CHP plant and the cost of electricity that may have to be bought in (as in the simple analysis described earlier in this section).

The "value" of the electricity is made up of the value of the electricity sold at the "buy-back" price ($(Y_E)_{BB}$ (£/kWh)) and the value of the electricity supplanted (i.e. formerly purchased from the grid, at a buy-in price $(Y_E)_{BI}$ (£/kWh)). Then, using the earlier notation ($\dot{W}_{CG}$ the power output of the CHP plant, and $\dot{W}_D$ the power demand) the value of the electricity is

$$H[(\dot{W})_{CG} - \dot{W}_D](Y_E)_{BB} + H\dot{W}_D(Y_E)_{BI} \quad \text{(£ p.a.)}$$

when H is the utilisation (hours per annum).

The cost of production is the fuel cost in the cogeneration plant less the savings obtained by stopping the separate generation of heat,

$$\left[\frac{\dot{W}_{CG} \mathscr{S}_{CG}}{\eta_{CG}} - \frac{\dot{Q}_D \mathscr{S}_B}{\eta_B} \right] H \quad \text{(£ p.a.)},$$

where $\mathscr{S}_{CG}$, $\mathscr{S}_B$ are the fuel prices (£/kWh (thermal)) in the new plant and the old boiler plant, no longer in use.

The money saved in the year is thus

$$A = [(\dot{W}_{CG} - \dot{W}_D)(Y_E)_{BB} + \dot{W}_D(Y_E)_{BI}]H - \left[\frac{\dot{W}_{CG}\mathscr{S}_{CG}}{\eta_{CG}} - \frac{(\dot{Q}_U)_D\mathscr{S}_B}{\eta_B}\right]H$$

$$= (Y_E)_{BI}\dot{W}_D H + \mathscr{S}_B(\dot{Q}_U)_D H/\eta_B - M_{CG}, \qquad (4.28)$$

where $M_{CG} = [\dot{W}_{CG}\mathscr{S}_{CG}/\eta_{CG} - (\dot{W}_{CG} - \dot{W}_D)(Y_E)_{BB}]H$. This is the cash flow (savings) given earlier in section 4.3, for $\dot{W}_{CG} > \dot{W}_D$, $(\dot{Q}_U)_{CG} = (\dot{Q}_U)_D$.

Porter and Mastanaiah also relate this cash flow to their concept of the incremental heat rate $(IHR)_{CG}$ determined from thermodynamic analysis. Thus instead of the term $(\dot{W}_{CG}/\eta_{CG})$ for the fuel energy requirement of the CHP plant, they use $\left[(IHR)_{CG}\dot{W}_{CG} + \frac{(\dot{Q}_U)_D}{(\eta_B)_{CG}}\right]$ as described in section 2.2.5. The net cash flow of equation (4.28) is then

$$A = [(\dot{W}_{CG} - \dot{W}_D)(Y_E)_{BB} + \dot{W}_D(Y_E)_{BI}]H$$

$$- \mathscr{S}_{CG}(IHR)_{CG}\dot{W}_{CG}H + (\dot{Q}_U)_D\left[\frac{\mathscr{S}_B}{\eta_B} - \frac{\mathscr{S}_{CG}}{(\eta_B)_{CG}}\right]H, \qquad (4.29)$$

where $(\eta_B)_{CG}$ is the boiler or combustion efficiency of the new plant.

If there is no internal demand $(\dot{W}_D = 0)$ and all the electricity $\dot{W}_{CG}$ is sold, then the cash flow is

$$A = \dot{W}_{CG}H[(Y_E)_{BB} - \mathscr{S}_{CG}(IHR)_{CG}] + (\dot{Q}_U)_D\left[\frac{\mathscr{S}_B}{\eta_B} - \frac{\mathscr{S}_{CG}}{(\eta_B)_{CG}}\right]H. \quad (4.30)$$

If no electricity is sold, and that generated fully supplants electricity previously bought from the grid, then $\dot{W}_{CG} = \dot{W}_D$. The cash flow is then

$$A = \dot{W}_D H[(Y_E)_{BI} - (\mathscr{S}_{CG}(IHR)_{CG}] + (\dot{Q}_U)_D\left[\frac{\mathscr{S}_B}{\eta_B} - \frac{\mathscr{S}_{CG}}{(\eta_B)_{CG}}\right]H. \quad (4.31)$$

The cash flows can also be corrected by allowing for increased capital costs, maintenance and operating costs for the CHP plant.

4.4 Examples of Economic Assessment Using DCF Techniques

4.4.1 Introduction

A full and detailed assessment of the economics of cogeneration was made by the Marshall Committee,[1] which reported to the UK Secretary of State for Energy in 1979. Marshall and his colleagues (the author included) made a particular study of combined heat and power for district heating (CHP/DH), considering

(i) *new* provision of commercial and domestic space and water heating for a population of 10,000 people—typical of a large town redevelopment or a new development in a suburb (heat load 20 MW);

(ii) *conversion* of a small city of 100,000 people to district heating, using CHP plant (heat load 200 MW);

(iii) *conversion* of a large city of 1,000,000 people to district heating, using CHP plant (heat load 2000 MW).

Marshall *et al*. made several basic assumptions (e.g. specifying the energy supply per "average" dwelling at 45 GJ per year) but varied several parameters such as the discount rate and the housing density. Their results are described in detail here.

Other numerical assessments of the economics of CHP schemes, also using DCF analysis, but applying criteria other than NPV, are also described later in this section.

4.4.2   The Marshall Study—Use of Simple Analysis for NPV

A simple algebraic analysis for net present value is first presented, based on the "change" type of approach. The money savings on fuel (obtained by changing from an existing fuel mix to a new heat and power scheme) are estimated, together with the extra capital costs involved. The net present value (NPV) of making the change (for an "average" dwelling) is then obtained, and presented in non-dimensional form. It is supposed that the utility is maintaining the overall supply of electricity (i.e. the sum of that required in the scheme and the demand external to the scheme).

In Chapter 3 we determined the fuel savings for making changes

(a) from an existing heating system, to a "heat only" central boiler system;

(b) from an existing (separate) heat and power system to a CHP "back-pressure" scheme;

(c) from an existing (separate) heat and power system to a CHP "pass-out" scheme.

We follow up those analyses here, to determine corresponding net present values, for (a), (b) and (c). A practical scheme (d) might involve building up the heat load with heat-only boilers (a) before bringing in a large "pass-out" CHP plant of scheme (c).

We first present the results in analytical form, giving a few numerical results. This simple analysis was given in an earlier report on district heating by the Marshall Group[11] (and given subsequently by Horlock and Owen[12]), but the Group's main and final report[1] presented more accurate results, based on computer calculations of discounted cash flow varying over several years, and these are described later. However, it was found[11,12] that the simple and complex calculations compared quite well, and it is instructive to present the analyses in the simple algebraic form before showing the more detailed results obtained from the Marshall computer program.

(a) Supplying the Domestic Heating Load by Central Boiler Plant

The first example of fuel saving considered in section 3.9.1 involved the replacement of electrical and domestic space heating loads by district heating from a central boiler. The total heat load $(\dot{\lambda})^*$ of a hypothetical average house, made up of an electrical space heating load $(\dot{W}_S)$, and domestic heating loads $(\dot{\lambda}_1, \dot{\lambda}_2)$ supplied by units of efficiency (η_{B_1}, η_{B_2}), were replaced by district heating from a central boiler of efficiency $(\eta_B)'$, and the heat demand changed to $\dot{\lambda}'$.

The total energy saving was found to be

$$\Delta \dot{F}_a = \frac{\dot{\lambda}_1}{\eta_{B_1}} + \frac{\dot{\lambda}}{\eta_{B_2}} + \frac{\dot{W}_S}{\eta_C} - \frac{(\dot{\lambda})'(1 + \phi)}{(\eta_B)'}, \qquad (4.32)$$

where ϕ was a loss factor allowing for transmission losses. This energy saving may be expressed in non-dimensional form as

$$\gamma_a = \left(\frac{\Delta \dot{F}}{\dot{\lambda}}\right)_a = \gamma_1 + \gamma_2 + \gamma_S - (\gamma_B)', \qquad (4.33)$$

or in terms of a fuel savings ratio based on the original energy supplied for heating the house

$$(\text{FESR})_a = 1 - \frac{(\gamma_B)' + \left(\dfrac{1}{\dot{\lambda}\eta_C}\right)}{\gamma_1 + \gamma_2 + \gamma_S + \left(\dfrac{1}{\dot{\lambda}\eta_C}\right)}; \qquad (4.34)$$

where

$$\gamma_1 = \frac{\dot{\lambda}_1}{\dot{\lambda}\eta_{B_1}}, \quad \gamma_2 = \frac{\dot{\lambda}_2}{\dot{\lambda}\eta_{B_2}},$$

$$\gamma_S = \frac{\dot{W}_S}{\dot{\lambda}\eta_C}, \quad (\gamma_B)' = \frac{(\dot{\lambda})'}{\dot{\lambda}}\left(\frac{1 + \phi}{(\eta_B)'}\right) = \frac{\delta(1 + \phi)}{(\eta_B)'},$$

and δ is the ratio of the new heat load $(\dot{\lambda})'$ to the old heat load $(\dot{\lambda})$. The annual money saving $(£\Delta M)$ associated with the energy saving is then

---

*Note that loads are expressed here as energy *rates* $(\dot{W}_S, \dot{\lambda}, \text{kW or GJ p.a.})$ so that dot superscripts are used, except for the total "external" power load which is unchanged and arbitrarily taken as unity.

obtained by dividing the individual terms in the energy saving equation (say in GJ p.a.) by calorific values (CV), GJ/kg, and multiplying by fuel prices (S, £/kg), i.e. multiplying by $\mathscr{S} = S/(CV)$. Thus

$$\Delta M = \frac{\mathscr{S}_1 \dot{\lambda}_1}{\eta_{B_1}} + \frac{\mathscr{S}_2 \dot{\lambda}_2}{\eta_{B_2}} + \frac{\mathscr{S}_C \dot{W}_S}{\eta_C} - \frac{(\mathscr{S}_B)'(\dot{\lambda})'(1 + \phi)}{(\eta_B)'}. \qquad (4.35)$$

Extra annual maintenance costs $\Delta(OM)$ may be subtracted from this saving due to fuel economy. A simple estimate of the extra capital cost is

$$\Delta C = (N + R) + \frac{c_B(\dot{\lambda})'(1 + \phi)}{(LF)} - c_C \dot{W}_S. \qquad (4.36)$$

The first term in brackets involves the extra capital of the distribution network (N) and the house internal costs (R). The second term represents the cost of the district heating boiler plant, of capital cost c_B (per unit heat rate) and with an overload factor (LF). The third term represents the capital saving (in an expanding economy) on *not* building a grid power station (of capital cost c_C per unit electrical power)) to supply the steady power load required for district heating. (Note that savings on the latter may be greater, as maximum load will be greater than $\dot{W}_S$.)

If we use the simplest estimate for the net present value of the saving (based on extra capital investment ΔC at year zero and immediate annual savings ΔM) then

$$(NPV)_a = \frac{\Delta M}{_N f_{AP}} - \Delta C. \qquad (4.37)$$

This may be expressed in non-dimensional form

$$(npv)_a = \frac{(NPV)_a}{\dot{\lambda} c_C} = m_1 \gamma_1 + m_2 \gamma_2 + m_C \gamma_S - (m_B)'(\gamma_B)'$$

$$- \left(\frac{c_N}{c_C} + \frac{c_R}{c_C} \right) - \frac{c_B \delta(1 + \phi)}{c_C(LF)} + \frac{\dot{W}_S}{\dot{\lambda}}, \qquad (4.38)$$

where $m = \dfrac{\mathscr{S}}{c_C(_N f_{AP})}$ is an operational factor of similar form to that first introduced by Baumann[13] (see Haywood[14]) and $c_N = \dfrac{N}{\dot{\lambda}}$, $c_R = \dfrac{R}{\dot{\lambda}}$ are the unit costs of the network and radiators.

(b) Supplying the Domestic Heating Load by a Back Pressure Turbine

In section 3.9.2 we also considered a change from an existing system to a CHP scheme using a back pressure steam turbine (the analysis is essentially the same for a gas turbine with waste heat recovery).

Noting that the power output and efficiency of the turbine are different from those of the central power station turbine it is replacing, the fuel saving was found to be

$$\Delta \dot{F}_b = \dot{F}_{REF} - \dot{F}_{CG}$$
$$= \left(\frac{\dot{\lambda}_1}{\eta_{B_1}} + \frac{\dot{\lambda}_2}{\eta_{B_2}} + \frac{\dot{W}_S}{\eta_C} \right) - \frac{(\dot{\lambda})'(1 + \phi)}{(1 - \eta_{CG})} \left(1 - \frac{\eta_{CG}}{\eta_C} \right). \quad (4.39)$$

In non-dimensional form

$$\gamma_b = \left(\frac{\Delta \dot{F}}{\dot{\lambda}} \right)_b = \gamma_1 + \gamma_2 + \gamma_S - \frac{\delta(1 + \phi) \left(1 - \dfrac{\eta_{CG}}{\eta_C} \right)}{(1 - \eta_{CG})}, \quad (4.40)$$

or

$$(FESR)'_b = \frac{\gamma_1 + \gamma_2 + \gamma_S - \dfrac{\delta(1 + \phi)}{(1 - \eta_{CG})} \left(1 - \dfrac{\eta_{CG}}{\eta_C} \right)}{\gamma_1 + \gamma_2 + \dfrac{1}{\lambda \eta_C} + \gamma_S}. \quad (4.41)$$

The extra capital cost is now

$$\Delta C_b = N + R + \Delta C_{CG} = N + R + [c_{CG}(1 - \dot{W}_C) + c_C \dot{W}_C] \\ - [c_C(1 + \dot{W}_S)], \quad (4.42)$$

where ΔC_{CG} is now the extra capital cost of building a special back pressure turbine. c_{CG} and c_C are the unit costs of cogeneration and conventional plant. The net present value, again based on the simplest analysis, is

$$(NPV)_b = \frac{\Delta M_b}{{}_N f_{AP}} - \Delta C_b$$
$$= \frac{1}{{}_N f_{AP}} \left[\frac{\mathscr{S}_1 \dot{\lambda}_1}{\eta_{B_1}} + \frac{\mathscr{S}_2 \dot{\lambda}_2}{\eta_{B_2}} + \frac{\mathscr{S}_C \dot{W}_S}{\eta_C} - \frac{\mathscr{S}_C \delta(1 + \phi) \dot{\lambda}}{(1 - \eta_{CG})} \left(1 - \frac{\eta_{CG}}{\eta_C} \right) \right]$$
$$- (N + R + \Delta C_{CG}), \quad (4.43)$$

or in non-dimensional form,

$$(\text{npv})_b = \frac{(\text{NPV})_b}{c_C \dot{\lambda}}$$

$$= m_1 \gamma_1 + m_2 \gamma_2 + m_C \gamma_S - \frac{m_C \delta (1 + \phi)}{(1 - \eta_{CG})} \left(1 - \frac{\eta_{CG}}{\eta_C} \right)$$

$$- \left(\frac{c_N}{c_C} + \frac{c_R}{c_C} + \frac{\Delta C_{CG}}{c_C \dot{\lambda}} \right). \tag{4.44}$$

(c) Replacement of Electrical and Domestic Heating Load by a Pass Out Turbine

The third example, considered in section 3.9.3, involved changing the existing system to a CHP scheme using a pass-out or steam extraction turbine, in which the ratio of lost power to district heating load was the factor Z encountered in the thermodynamic analysis of section 2.6.2. The fuel saving was shown to be

$$\Delta \dot{F}_c = \dot{F}_{REF} - \dot{F}_{CG} = \frac{\dot{\lambda}_1}{\eta_{B_1}} + \frac{\dot{\lambda}_2}{\eta_{B_2}} + \frac{\dot{W}_S}{\eta_C} - \frac{Z \delta (1 + \phi)(\dot{\lambda})}{\eta_C}. \tag{4.45}$$

In non-dimensional form

$$\gamma_c = \gamma_1 + \gamma_2 + \gamma_S - \frac{Z \delta (1 + \phi)}{\eta_C}, \tag{4.46}$$

or

$$(\text{FESR})_c = \frac{\gamma_1 + \gamma_2 + \gamma_S - Z \delta (1 + \phi)/\eta_C}{\gamma_1 + \gamma_2 + \gamma_S + (1/\dot{\lambda}\eta_C)}. \tag{4.47}$$

The extra capital cost is now

$$\Delta C_c = N + R + \Delta C_{CG}, \tag{4.48}$$

where $\Delta C_{CG} = [c_{CG}(1 - \dot{W}_C) + c_C \dot{W}_C] - [c_C(1 + \dot{W}_S)]$ is the extra capital required to build the pass-out turbine compared with a standard turbine. Note that power output $\dot{W}_C = Z \delta \dot{\lambda}(1 + \phi)$ has been lost and extra capital is required to build plant of efficiency η_C elsewhere, but a credit is again available since overall electrical demand has been reduced by the elimination of electrical space heating ($\dot{W}_S$).

The net present value is

$$(\text{NPV})_c = \frac{\Delta M_c}{N f_{AP}} - \Delta C_c$$

$$= \frac{1}{N f_{AP}} \left[\frac{\mathscr{S}_1 \dot{\lambda}_1}{\eta_{B_1}} + \frac{\mathscr{S}_2 \dot{\lambda}_2}{\eta_{B_2}} + \frac{\mathscr{S}_C \dot{W}_S}{\eta_C} - \frac{\mathscr{S}_C Z \delta (1 + \phi) \dot{\lambda}}{\eta_C} \right]$$

$$- [N + R + \Delta C_{CG}], \tag{4.49}$$

or in non-dimensional form,

$$(\text{npv})_c = m_1 \gamma_1 + m_2 \gamma_2 + m_C \gamma_S - m_C Z \delta (1 + \phi)/\eta_C$$

$$- \left[\frac{c_N}{c_C} + \frac{c_R}{c_C} + \frac{\Delta C_{CG}}{c_C \dot{\lambda}} \right]. \tag{4.50}$$

(d)   Two-stage Conversion

If we consider a two-stage operation involving initially conversion from the existing fuel supply to a heat only boiler (a) and a subsequent move (d) to CHP using a pass-out turbine, then the non-dimensional net present value of this second stage (d) is

$$(\text{npv})_d = (m_B)'(\gamma_B)' - \frac{m_C Z \delta (1 + \phi)}{\eta_C} - \left(\frac{\Delta C_{CG}}{c_C \dot{\lambda}} - \frac{\dot{W}_S}{\dot{\lambda}} \right), \tag{4.51}$$

assuming that the heat only boilers are not used again at the end of the first stage. The full conversion (c), from an existing scheme to a pass-out turbine, is made up of (a) plus (d), i.e.

$$(\text{npv})_c = (\text{npv})_a + (\text{npv})_d. \tag{4.52}$$

Numerical Examples

Horlock and Owen[11] calculated a numerical example of the conversion of a large city in these two stages ((a) plus (d) equals (c)) using the algebraic expressions derived. The results were presented in non-dimensional form; the approximate 1976 prices used were broadly consistent with those given in Table 4.3 below.

Figure 4.4 shows the various components making up $(\text{npv})_a$ and $(\text{npv})_d$, and hence the total net present value of the two stage conversion $(\text{npv})_c$. The left-hand side of the figure illustrates the first stage conversion to heat only boilers (a) and shows a net *cost* resulting. Savings on fuel costs are substantial, together with the saving on *not* building a power station to meet the electrical space heating demand. But there are substantial capital costs involved in the network and house modification (installation of

TABLE 4.3. Some Assumptions made in the Marshall Report

1. Fuel Price ($\mathscr{P}$)

	p/GJ
Gas—District Heating Boiler	x
Residual oil or coal—District Heating Boiler	x − 10a
Gas Oil—District Heating Boiler	x
Oil or Gas—Domestic Heating Boiler	x + 35a
Residual Oil or Coal—Large Power Stations	x − 35a
Residual Oil—Small Power Stations	x − 10a

"x" was given the value of 130 p/GJ for a constant (in real terms) case (1976 prices), and 240 p/GJ as an equivalent for the doubling in real terms, (for a scheme introduced in 1985 discounted at 10% over the life of the scheme).

The sensitivity to relative price differences was examined by varying "a" from 0.5 to 2.0 with a central value of 1.0.

2. Fuel Mix

Type of Heating	% of dwellings	Energy delivered (GJ/Dwelling/Yr)
Electricity	20	43
Wet central heating	30	69
Other	50	79
Total/Average	100	69

3. Heating Requirements

Heating System	Delivered energy requirement (GJ/Dwelling/Year)		
	Space	Water	Total
Modern gas central heating (separator water heater)	45	15	60
On-peak electric heating	30	12	42
Off-peak electric heating	33	12	45
District heating	30	15	45

4. Initial Domestic Heating Capital Costs

£

Gas Fired Central Heating

Initial Cost of Installation	= 750	
Radiator replacement—£100 at 10-year intervals, present worth	= 63	
Boiler replacement —£150 at 20-year intervals, present worth	= 26	
Maintenance at £16/year present worth	= 159	
	998	

District Heating

Initial cost of installation	500	
Radiator replacement—£100 at 10-year intervals, present worth	= 63	
	563	

5. Plant Assumptions

(i) Costs of Plant

Station Type	Station size MW (e)	Capital costs £/kW (e)	Running costs (fuel)	Fixed operating costs £/kW (e)	Life (years)	Electricity "loss" per unit heat (Z)
Large fossil	2000	315	$(x - 35a)$ p/GJ (35%) effy)	30	30	0.14
Small Back Pressure	20	440	$(x - 10a)$ p/GJ (17.1% effy)	105	30	0.16

(ii) Details of CHP Schemes Studied

	Green Field	Small City	Large City
Heat Load supplied	20 MW	200 MW	2000 MW
Plant	2 small back pressure sets	2 × 500 MW (e) (large fossil)	4 × 500 MW (e) (large fossil)
Turbine Conversion/Modification	Purpose built	Multi-stage extraction pass out conversion	Multi-stage extraction pass out conversion
Cost of Conversion/Modification			
(a) Replacement of plant for lost electrical capacity	—	£42.5/kW (heat supplied) £200/dwelling	— £200/dwelling
(b) Turbine modifications	—	£5/kW (heat supplied) £42/dwelling	— £42/dwelling
(c) Outage costs while converting sets	—	£3/kW (heat supplied) £84/dwelling	— £17/dwelling
No. of sets installed	2	2 (1 standby)	4 (1 standby)

6. Transmission Cost (CHP)

	Green Field	Small City	Large City
Transmission System			
Length installed	None	15 km	15 km
Capital Cost		£1000/m/m (+20% for urban working)	
Number of Pipes		Single pair of go and return pipes	
Load factor		50%	
Temperatures		120°C (supply), 60°C (return)	
Pumping Power Costs		2.5 p/kWh	
Flow Velocity		2.8 m/s	
Primary Distribution System			
System Installed	None	Yes	Yes
Capital Cost		£1000 $K^{\frac{1}{3}}$ per dwelling (K = load in MW/km^2)	
Secondary Distribution System			
System Installed	Yes	Yes	Yes
Capital Cost	£$(9 + 1000\ K_0/K)$ per kW where $K_0 = 1$ MW/km^2, K = actual heat density of the area	+50%	+50%
Factor on capital for interest during build-up	1.38	1.55	1.55

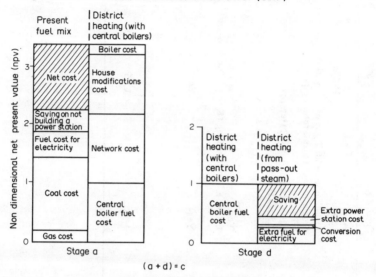

FIG. 4.4. Non-dimensional net present value (npv) for two stage
conversion to district heating.

radiators, etc.) and these more than outweigh money savings on fuel
economy in the net present value. Note that the costs of the central boiler
are relatively small.

In the second stage of conversion (d), shown on the right-hand side of
Fig. 4.4 there are substantial savings on $(npv)_d$. The extra fuel and capital
costs of the extraction turbine are small compared with the contribution of
fuel savings in the heat only boiler.

However, as the numerical calculations showed, the savings on $(npv)_d$
are not sufficient to balance the net cost in stage (a), and there is an overall
net *cost* for (c) ((a) plus (d)), when expressed in terms of $(npv)_c$ for this
particular example.

These calculations, based on the simple "algebraic" analysis, are illus-
trative only, and use approximate 1976 prices (assumed in 1978 when
Energy Paper 20[11] was written). More accurate calculations, undertaken
by the ETSU Group at Harwell for the Marshall Committee, involved
more complex money distributions than that assumed in the simplest
analysis (single capital investment and immediate money savings) but used
the same prices. Although allowance for particular distributions of investment with time, and corresponding money savings, can also be made in the
algebraic analyses (see Appendix B), the ETSU computer study (by Hewitt
and Owen) for the Marshall report[1] was also able to include such effects
as provision for overload, stand-by operation, non-availability of plant,

etc. The results of these more detailed calculations are considered in the next section.

4.4.3   The Marshall Study—Detailed Computer Calculations

The extensive results computed for the original Marshall study are now reviewed. It is important first to state the assumptions that were used in the calculation, together with the parameters that were varied.

As explained earlier, the calculations were made for a city of 10,000 people on a "green-field" site; a medium sized existing city of 100,000 population: and a large city of 1,000,000 population.

Further the analyses were made for

(i) the medium term (up to the year AD 2000) in which it was assumed that natural gas was available as the main competitor;

(ii) the longer term (after the year AD 2000) when CHP/DH would have to compete with electrical heating, synthesized natural gas, heat pumps, and coal-fired heating.

It is the results calculated for the medium term that are presented here. The full set of assumptions are given in ref. 1, but important ones used in the medium term study are reproduced in Table 4.3.

Even this abbreviated listing of the assumptions made illustrates the complexity of the calculations. Costs used were those at January 1976 levels and lifetime costs were discounted over 60 years. Following Treasury guidelines the principal value of the discount rate was taken as 10%, it being assumed that inputs and outputs were not affected by changes in general price levels, but allowance was made for changes in *relative* price levels e.g. fuel costs (i.e. an "effective" discount rate allowing for inflation (as used by Williams[2]) was not used on Treasury advice).

The method of analysis was essentially the simple algebraic approach described in section 4.4.2, but instead of savings being estimated in a "change" type of analysis, the costs per dwelling with CHP/DH were compared directly with those for (i) an existing fuel mix (for an "average" dwelling) and (ii) gas-fired central heating. A levelised annual cost (£/dwelling year)

$$\bar{A} = {}_\text{N}f_\text{AP}(\text{NPV}) \tag{4.27}$$

(see Appendix B) was derived for CHP/DH compared with (i) and (ii). The detailed calculations allow for the cash flows to vary year by year as the plant and the savings build-up.

The important parameters varied in the Marshall report were

(i) housing density;
(ii) discount rate;
(iii) fuel prices, and their rate of change.

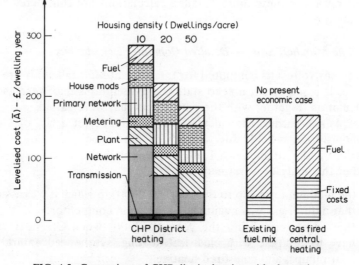

FIG. 4.5. Comparison of CHP district heating with alternatives
(after Marshall[11]). Large city (population 1,000,000).
Assumptions 10% discount rate; present fuel prices.

Housing density has a major effect on the capital cost of transmission (in
the secondary distribution system), high density of the heat load (HDHL)
reducing such costs sharply. Discount rate clearly affects the capital
charges (and hence both NPV and $\bar{A}$). Relative fuel prices have a major
influence on the difference between fuel costs in CHP/DH and the system
it replaces (and hence both the NPV and $\bar{A}$).

The effects of these parameters on levelised cost ($\bar{A}$) are clearly shown
in Figs. 4.5 and 4.6, for the large city of 1,000,000 people. Figure 4.5
indicates that with "present" (1976) fuel prices and the recommended
10% discount rate, there is no economic case for CHP/DH, even for the
highest density housing (50 dwellings per acre). Network costs drop rapidly
with housing density, but not sufficiently to give CHP/DH any advantage
over either the existing fuel mix or gas-fired central heating.

Figure 4.6 shows a different picture, when the discount rate is reduced
to 5%, and fuel prices are assumed to double by AD 2000. Now the fuel
costs become dominant for both (i) the existing fuel mix and (ii) gas-fired
central heating, while the capital costs associated with CHP/DH are
greatly reduced.

Broadly similar results were obtained for the green-field site and the
medium sized city, where the CHP/DH power plants were assumed to be
two small back pressure sets for the first example and a multi-stage steam
extraction plant for the second.

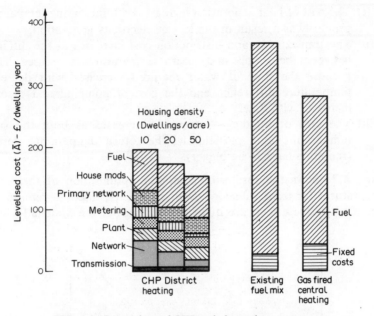

FIG. 4.6. Comparison of CHP and alternatives—a more optimistic view (after Marshall[11]). Large city (population 1,000,000). *Assumptions* 5% discount rate; fuel prices doubling by 2000.

4.4.4 Subsequent Calculations

Critical assumptions originally made in the Marshall calculations were

(i) that the temperature difference between water supply (at 120°C) and water return (60°C) was constant;

(ii) that large pass-out sets were used for the two city cases rather than back-pressure sets;

(iii) that heat loads were uniformly distributed in the city areas, with no "holes" in the network (due to parks and unconnected areas).

A subsequent report by Macadam *et al.*[15] studied these assumptions in more detail, together with others which proved less critical (the effect of metering, and the cost effectiveness of improved insulation). Their results (on the three critical assumptions only) are given below.

4.4.4.1 Optimisation of Distribution Temperatures

There are three major effects of changing district heating water temperatures.

(i) The cost of heat generation increases with increasing temperature, primarily as a result of the loss of electricity generation.

(ii) The transmission and distribution cost increases as the difference between the supply and return temperatures decreases. This is because the flow of water required increases with decreasing temperature difference and the cost of pumping and pipework increases with flow.

(iii) The cost of building "internals" increases as both the supply temperature and, particularly, the return temperature decrease (radiators have to be bigger).

Figure 4.7 shows the results of computer optimisation of these water temperatures for fixed values of dwelling density, discount rate and fuel price. Clearly the effects are quite substantial, but the Marshall assump-

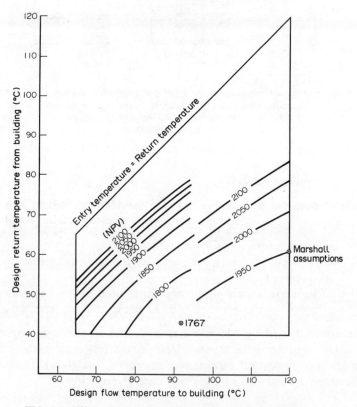

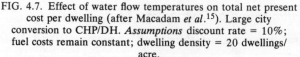

FIG. 4.7. Effect of water flow temperatures on total net present cost per dwelling (after Macadam *et al.*[15]). Large city conversion to CHP/DH. *Assumptions* discount rate = 10%; fuel costs remain constant; dwelling density = 20 dwellings/ acre.

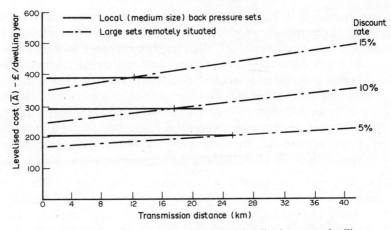

FIG. 4.8. Effect of transmission distance on levelised cost per dwelling (after Macadam *et al.*[15]). (Fuel prices rise exponentially, doubling by the year 2000).

tions were fortunately quite close to the optimum and the original results were not invalidated.

4.4.4.2 Effect of Using Local (Medium Sized) Back Pressure Sets

The original Marshall calculations for a large city were based on 2000 MW pass out sets and a typical transmission distance of 15 km. Macadam *et al.*[15] considered the alternative of using smaller, locally sited back-pressure sets (with lower efficiency, but less transmission costs and losses). Figure 4.8 shows the results of this second series of calculations. The small sets provide slightly lower overall costs per dwelling than the larger sets, when those larger sets are located far from the heat load. The "cross-over" point (the transmission distance above which the large sets become less economic than the small sets) varies with discount rate, from 12 km at 15% to 24 km at 5%.

4.4.4.3 Network Costs—Non-Uniformities in Heat Load

The Marshall study[1] assumed heat loads to be uniformly distributed throughout city areas, and applied the formula given in Table 4.3 to obtain (secondary) network costs. While it allows for overall (averaged) heat loading, this formula does not specifically include the effect of the number of demand points, or their distribution, as Cassels[16] points out.

By studying a series of model cities with various geometries of heating distribution (with radially varying load and different patterns of heated and unheated areas), Macadam et al.[15] calculated the network cost for a typical large city. They concluded that the Marshall studies had overestimated network costs by a factor of about 2 (leading to heating costs some 15% less than those given in ref. 1).

These estimates of network costs are complex and the reader is referred to the various studies of the problem (Appendix 7 of ref. 11 and refs. 15, 16 and 17). They are also very important for, as Fig. 4.5 shows, network costs have a major effect on CHP economics, particularly at high discount rates.

4.4.4.4 Heat Loads in Cities

The calculations undertaken by the Marshall group, and reported initially in ref. 11, showed the dominant effect of housing density (through the network costs discussed above).

A further report[17] published in parallel with ref. 1, assessed the amount of "high density heat load" (the heat load with a minimum density of 20 MW/km^2) in Great Britain, and found it to be about 32.2 GW. Most of this high density heat load (79%) was found to be in the five largest conurbations. The study enabled estimates to be made of the energy savings that would follow from connecting the heat load to CHP systems, and concluded that compared with the present systems substantial savings would be made (see the Introduction to this book). This was an important result, for it gave a realistic estimate of energy savings that would follow the installation of an amount of cogeneration plant that was itself economic (in cities with heat loading densities sufficiently high to keep capital costs down).

4.4.5 Other Studies

Many other studies have been made of the economics of CHP schemes using DCF techniques. Here we refer to a few of these to illustrate the wide variety of such studies; classification is on the basis of the criteria used in the various assessments.

4.4.5.1 Investor's Rate of Return (ROI)

Reference has already been made to the work of Daudet and Trimble,[7] who used ROI (the investor's rate of return) as the critical deciding criterion. They determine the net present value (or worth) as in the Marshall studies, i.e.

$$\text{NPV} = \sum_{k=1}^{N} \frac{A_k}{(1 + i)^k}. \tag{B.10}$$

They adopt a "change" type of analysis, interpreting A (the net cash flow) as the difference between the costs of an (old) existing scheme (electricity purchased, fuel costs, maintenance and capital costs) and those of the (new) cogeneration scheme. Allowance is made for income tax and depreciation. The ROI is that discount rate which sets the NPV at zero over a plant life N.

However, in this determination of ROI, a critical point is whether the capital needed for the CHP plant is supplied internally (by the institution installing the plant) or raised externally. Daudet and Trimble[7] point out that if the fraction of the extra capital raised externally is z, and the loan repayment associated with that capital $(z\Delta C)$ is ΔA_{LRP} then the net cash flow (savings) is reduced by this amount,

$$A' = A - \Delta A_{\text{LRP}}. \tag{4.53}$$

It is this cash flow year by year which has to be discounted and compared with the "internal" capital used $(1 - z)\Delta C$. The simple DCF analysis of section 4.3.1 is then modified to give the net present value as

$$\text{NPV} = \frac{A'}{{}_{\text{N}}f_{\text{AP}}} - (1 - z)\Delta C. \tag{4.54}$$

Thus the company's rate of return is given by the discount rate which satisfies the equation $(\text{NPV}) = 0$, i.e.

$$(1 - z)\Delta C = \frac{A'}{{}_{\text{N}}f_{\text{AP}}}. \tag{4.55}$$

A similar approach using ROI is used by Flint and El-Masri,[9] who made a careful study of its sensitivity to pricing systems, tax and financial variables for CHP schemes at four hospital sites in the United States. There are two points of particular interest in their work.

(a) They consider the annual variation in heat and electrical load (see, for example, Fig. 4.9 for a hospital at San Diego) and then suppose the overall demand is met by
   (i) a cogeneration plant supplying constant base loads of heat and electricity (the dotted lines in Fig. 4.9), and
   (ii) bought-in heat or electricity to meet the supplementary requirements.
   The cash flows therefore have to take account of these purchases.

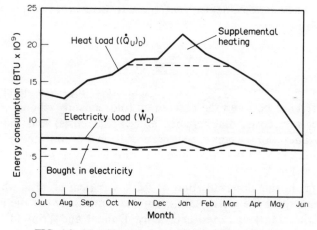

FIG. 4.9. Monthly energy use for San Diego hospital
(after Flink and El-Masri[9]).
$(BTU \times 10^9 \equiv MJ \times 10^6 \times 1.055)$

(b) A range of pricing and system variables were studied, i.e.
   capital cost;
   gas price;
   operational and maintenance costs;
   standby charges;
   inflation;
   turbine size;
   electricity price;
   turbine efficiency.

In addition the following tax and financial parameters were varied:

   depreciation life;
   interest rate;
   investment tax credit;
   repayment period;
   borrowed capital.

Nominal values of both sets of parameters were selected and the ROI for the four sites determined (see Table 4.4).

   The San Diego hospital scheme is obviously the most attractive (largely because of high local electricity prices). The tax levels and the fraction of capital borrowed (z) clearly have major effects. The sensitivity of ROI to variation of pricing and system parameters from nominal values is shown in Fig. 4.10(a), and to variation of tax and financing parameters in Fig. 4.10(b), both for the San Diego hospital. Calculations for a hospital at

TABLE 4.4. *ROI (%) For The Four Sites Considered Based on The Nominal Values of the Parameters*

Capital Borrowed (z)	Tax Exempt		After Tax	
	0	0.5	0	0.5
Buffalo, N.Y.	41.3	64	33.9	46.1
San Diego, Ca.	89.2	158.2	65.8	104.4
Lake City, Fl.	23.2	30.7	20.3	23.4
Battle Creek, Mi.	26.5	36.5	22.9	27.6

Key
1 Capital cost
2 Gas price
3 O&M costs
4 Stand by charges
5 Inflation
6 Turbine size
7 Electricity price $((Y_E)_{BI})$
8 Turbine efficiency

Key
A Depreciation life
B Interest rate
C Investment tax credit
D Repayment period
E % Borrowed capital

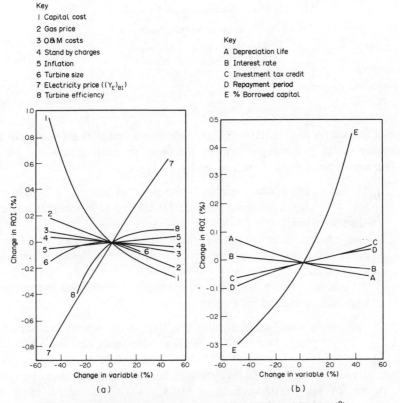

FIG. 4.10. San Diego hospital scheme (after Flink and El-Masri[9]).
(a) Sensitivity of ROI to system variables (San Diego) (0% borrowed capital). (b) Sensitivity of ROI to financial variables (San Diego) (50% borrowed capital).

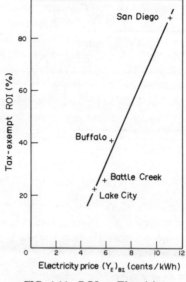

FIG. 4.11. ROI vs. Electricity
price (Flink and El-Masri[9]).

Lake City, where electricity prices are less, show that in the assessment of financing variables the repayment period becomes more critical.

The conclusions from these sensitivity analyses are as follows:

(i) The most important variable is the price of electricity bought in; Fig. 4.11 shows the tax-exempt ROI for the four schemes against electricity price.

(ii) one of the least sensitive parameters is the thermodynamic efficiency of the gas turbine in the CHP plant, unless it is reduced very substantially (i.e. well below 20%);

(iii) repayment period and the percentage of borrowed capital are important financial parameters.

Porter and Mastanaiah[8] similarly use the ROI as an assessment criteria, but for a larger plant. Sensitivity of ROI to electricity costs was demonstrated, for a plant of demand 10 MW (e) and 50 MW (th). Coal-fired steam turbines were found to be optimum for particular sale and "bought-in" prices ($(Y_E)_{BB} = 0.034$, $(Y_E)_{BI} = 0.044$, \$/kWh). However, doubling of electricity values shifted the optimal case to a large gas-fired gas turbine.

4.4.5.2   Break-even Capital Cost (BECC)

Jody *et al.*[6] use the break-even capital cost (BECC) as the deciding criterion for the economics of CHP schemes. However, as has already been pointed out, the application of this criterion again rests on the determination of the net present value, as in the Marshall studies.

Jody *et al.* determine the BECC by assessing the capital investment that sets the NPV to zero for a given plant life and a specified discount rate. They study the BECC for several small CHP/IND schemes with a "demand" heat/power ratio (λ_D) equal to 1.66, and the heat load in the form of a saturated steam supply of 110 psia (0.758 MPa).

The potential cogeneration schemes included

(i) a gas turbine (with a waste heat boiler);
(ii) a natural gas-fired non-condensing steam turbine;
(iii) a coal-fired non-condensing steam turbine;
(iv) Diesel engines with a waste heat boiler.

and their characteristics are given in Table 4.5.

The basic assumptions made by Jody *et al.* in their studies of CHP/IND were as follows:

on-site boilers, gas turbines and natural gas-fired turbines all use natural gas priced at $2.80/10^6$ Btu ($2.65/GJ);

Diesel systems use Diesel fuel priced at $7.00/10^6$ Btu ($6.63/GJ);

coal-fired steam turbines use coal at $1.90/10^6$ Btu (delivered) ($1.80/GJ);

industrial return on investment = 20%;

required return on investment on high risk ventures = 30% (before taxes);

electric energy market value = $0.045/kWh;

cost of maintenance of the cogeneration systems = $0.0012/kWh;

escalation rate for the cost of maintenance = 7% annually;

TABLE 4.5. *Characteristics of Potential Cogenerators*

System	Work/Heat ($1/\lambda_{CG}$)	Fraction of Heat Usable by Process
1. Gas Turbine	0.40	0.60
2. Natural Gas-fired Non-Condensing Steam Turbine	0.20	0.80
3. Coal-fired Non-Condensing Steam Turbine	0.20	0.80
4. Diesel Engines	0.57	0.30

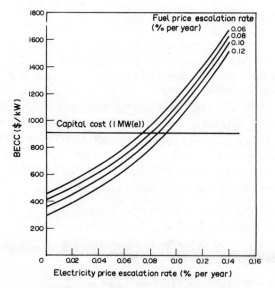

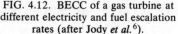

FIG. 4.12. BECC of a gas turbine at
different electricity and fuel escalation
rates (after Jody *et al.*[6]).

depreciation rate = 10% annually;
cogenerator life = 20 years;
hours of operation per year = 8760;
investment tax credit = 10%;
taxes = 50%;
all costs in 1980 dollars.

Figures 4.12 and 4.13 show the BECC for two of the plants studied (the
gas turbine and the coal-fired non-condensing turbine). The variables are
the electricity and fuel escalation rates (percentage price increase per
year).

Where the BECC is greater than the probable capital cost of the plant
(e.g. for the gas turbine with high electricity escalation rates) a CHP
scheme is economic. The required rate of return will certainly be met by
the probable capital spent, if not bettered. Figure 4.12 shows that if electricity prices are held constant (a vertical line) and fuel prices increase, the
CHP scheme becomes less feasible. While this is correct (because of
increased cost of running the CHP plant) it is unlikely that electricity costs
from outside utilities could be held constant while fuel prices rose.

Jody *et al.* conclude from these studies that for small plant (1 MW (e))
steam turbines and Diesel plant are not cost effective, but gas turbines
using natural gas have the potential to be cost effective.

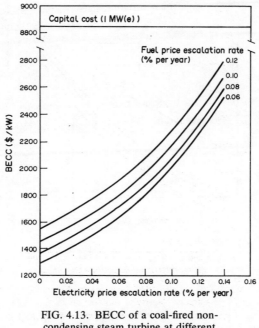

FIG. 4.13. BECC of a coal-fired non-
condensing steam turbine at different
electricity and fuel escalation rates
(after Jody *et al.*[6]).

4.5   Unmatched Operation of CHP Plant "Off-Design"–Effect on Economics

Both Chapter 3 (section 3.4) and several of the sections in this chapter have referred to the design-point operation of CHP plant under "unmatched" conditions, i.e. when the heat and power demands are not matched exactly to the heat and power supplied by the CHP plant; extra heat and or electricity has to be bought, or excess electricity sold. This aspect is discussed further in the section but relative to "off-design" performance of a given plant; in particular the work of Lilley[5] is developed.

4.5.1   Limits on Design Point Operation

The heat and electrical power *output* from a particular CHP plant may be plotted for a range of operating conditions (see Fig. 4.14(a)). Ideally the heat and electrical power demands $((\dot{Q}_U)_D, \dot{W}_D)$ are met by continuous operation at the design point $(\dot{Q}_U{}^*, \dot{W}^*)$, i.e. $(\dot{Q}_U)_D = \dot{Q}_U{}^*, \dot{W}_D = \dot{W}^*$.

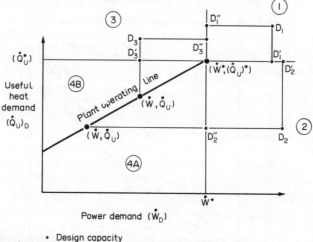

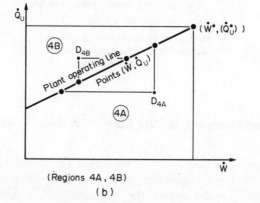

FIG. 4.14. Off design operation of CHP plant.

As discussed in section 4.3.1, the costs of running the original plant for H hours per annum will be

$$M_C = \left[(Y_E)_{BI} \dot{W}_D + \frac{\mathscr{S}_B (\dot{Q}_U)_D}{\eta_B} \right] H + (OM)_C, \tag{4.56}$$

where $(Y_E)_{BI}$ and $\mathscr{S}_B$ are the "bought-in" electricity price and the price of fuel for the boiler, respectively. The cost of operating the replacement cogeneration (CHP) plant, ideally matched, will be

$$M_{CG} = \frac{\dot{W}_D \mathscr{S}_{CG} H}{\eta_{CG}} + (OM)_{CG}, \tag{4.57}$$

where η_{CG} is the efficiency of the CHP plant, $\mathscr{S}_{CG}$ is the price of the fuel used in the plant, and $(OM)_{CG}$ the (increased) operational and maintenance costs per annum. Lilley points out that an immediate limit on the feasibility of CHP schemes is imposed by the condition that $(M)_{CG}$ must be less than M_C, i.e. that

$$\frac{\mathscr{S}_{CG}}{\eta_{CG}} + \frac{\Delta(OM)}{\dot{W}_D H} < (Y_E)_{BI} + \frac{\mathscr{S}_B \lambda_D}{\eta_B}, \qquad (4.58)$$

where $\lambda_D = (\dot{Q}_U)_D / \dot{W}_D$, $\Delta(OM) = (OM)_{CG} - (OM)_C$.

If this inequality cannot be satisfied for this ideally matched CHP plant, meeting electrical and heat demands exactly, there is no point in considering the matter further through a complex DCF analysis.

For example, consider a CHP gas turbine operating with a continuous gas supply priced at £9.2 per MWh and with a thermal efficiency of 20%, compared with 80% efficient boilers operating with the same fuel, and electricity priced at £20/MWh. Suppose further that the energy utilisation factor of the turbine is

$$\text{EUF} = \frac{\dot{W}_D + (\dot{Q}_U)_D}{\dot{F}_D} = 0.70,$$

$$\text{so that} \frac{(\dot{Q}_U)_D}{\dot{F}_D} = 0.50, \quad \text{and} \quad \lambda_D = \frac{(\dot{Q}_U)_D}{\dot{W}_D} = 2.5.$$

Thus, referring to the inequality (4.58), the right-hand side is

$$(Y_E)_{BI} + \frac{\mathscr{S}_B \lambda_D}{\eta_B} = 20 + \left(\frac{9.2 \cdot 2.5}{0.80}\right) = \text{£48.75/MWh},$$

and the left-hand side is

$$\frac{\mathscr{S}_{CG}}{\eta_{CG}} + \frac{\Delta(OM)}{\dot{W}H} = \frac{9.2}{0.2} + 1 = \text{£47/MWh},$$

if the increased costs of maintenance are £1/MWh. The scheme *may* thus be just viable. However, if the gas to the original boilers were priced on a different tariff (say on an "interruptable" basis at £7.20 per MWh) then the right-hand side of the inequality would be

$$20 + \left(\frac{7.5 \cdot 2.5}{0.8}\right) = \text{£42.5/MWh},$$

and there would be little point in pursuing the proposal further.

4.5.2 Types of Off-Design Operation

The operation of the CHP plant off-design, with the electrical power and/or the heat demand $(\dot{W}_D, (\dot{Q}_U)_D)$ not equal to that delivered by the plant at its design point, is complex. Referring to Fig. 4.14(a), the demand locus may be located in any one of the four quadrants 1, 2, 3, 4 relative to the design operating condition of the plant $((\dot{W})^*, (\dot{Q}_U)^*)$. Further, the plant may continue to be operated at its design point or "off-design" $(\dot{W} < (\dot{W})^*, \dot{Q}_U < (\dot{Q}_U)^*)$ at points along its operating line. We consider below possible modes of operation for the demand point located within each of the four quadrants.

Quadrant 1

In this case both extra heat $[(\dot{Q}_U)_D - (\dot{Q}_U)^*]$ and extra electricity $(\dot{W}_D - (\dot{W})^*)$ have to be bought in, even when the plant is operating at its maximum power output. This is really a somewhat unrealistic case, since the plant has clearly not been sized to match maximum demand of heat or electricity (although the hospital plants described in section 4.4.5 were operated this way). However, two limiting cases are more likely. In the first case D_1' $((\dot{Q}_U)_D = (\dot{Q}_U)^*, \dot{W}_D > (\dot{W})^*)$ electricity only needs to be bought in. In the second case D_1'' $((\dot{Q}_U)_D > (\dot{Q}_U)^*, \dot{W}_D = (\dot{W})^*)$ extra heat has to be purchased via heat only boilers.

Quadrant 2

The plant may be operated at its design point $(\dot{W})^*, (\dot{Q}_U)^*$ and we consider this option first. In this case excess heat $(\dot{Q}_U)^* - (\dot{Q}_U)_D$ is available for sale, but since electrical demand is greater than that generated, electricity $[\dot{W}_D - (\dot{W})^*]$ has to be bought in. Alternatively, the plant may be operated off-design at the point $(\dot{W}, \dot{Q}_U)$ on the operating line, matching the heat demand with $\dot{Q}_U = (\dot{Q}_U)_D$. No excess heat is then available for sale, but more electricity $(\dot{W}_D - \dot{W})$ has to be bought in.

Limiting cases in this quadrant are D_2', D_2'', for which $(\dot{Q}_U)_D = (\dot{Q}_U)^*$, $\dot{W}_D > (\dot{W})^*$ and $\dot{W}_D = (\dot{W})^*$, $(\dot{Q}_U)_D < (\dot{Q}_U)^*$ respectively.

Quadrant 3

If the plant is operated at its design point $(\dot{W})^*, (\dot{Q}_U)^*$ then surplus electricity $[(\dot{W})^* - \dot{W}_D]$ may be sold, but extra heat $[(\dot{Q}_U)_D - (\dot{Q}_U)^*]$ has to be bought in. Alternatively the plant may be operated off-design at $(\dot{W}, \dot{Q}_U)$, with the electrical demand matched $(\dot{W} = \dot{W}_D)$ but extra heat bought in.

Limiting cases in this quadrant D_3', D_3'', for which $(\dot{Q}_U)_D = (\dot{Q}_U)^*$, $\dot{W}_D < (\dot{W})^*$ and $\dot{W}_D = (\dot{W})^*$, $(\dot{Q}_U)_D > (\dot{Q}_U)^*$ respectively.

Quadrant 4

Operation in Quadrant 4 may be divided into two further operational sub-regions, 4A and 4B, below and above the "off-design" line of the plant (Fig. 4.14(b)).

In the region 4A demand for electricity ($\dot{W}_D$) and heat ($\dot{Q}_U)_D$ are both less than can be supplied by the plant at its design point. Thus operation of the plant at its design point $(\dot{W})^*$, $(\dot{Q}_U)^*$ would mean that surplus electricity and heat were for sale. It would be more usual to match the electrical demand by operating the plant off-design at $\dot{W}, \dot{Q}_U$ (such that $\dot{W} = \dot{W}_D$) with surplus heat available for sale. Alternatively, but probably less attractive economically, the heat load could be matched by operation at $(\dot{W}, \dot{Q}_U)$ with $(\dot{Q}_U) = (\dot{Q}_U)_D$. In this case extra electricity $(\dot{W}_D - \dot{W})$ would have to be bought in. Thus in terms of the three options: for operation at $[\dot{W}^*, (\dot{Q}_U)^*]$ both electricity and heat may be sold; for operation off-design at $(\dot{W}, \dot{Q}_U)$ heat may be sold, or electricity must be bought in.

In region 4B, demand for heat and electricity is again less than that which can be supplied by the plant at design point operation, so if the plant is run at $[(\dot{W})^*, (\dot{Q}_U)^*]$ surplus heat and electricity may both be sold. Alternatively heat demand may be matched by operating at $(\dot{W}, \dot{Q}_U)$ with $\dot{Q}_U = (\dot{Q}_U)_D$ and surplus electricity can still be sold. However, if the electricity demand is matched by operation off-design at $(\dot{W}, \dot{Q}_U)$, with $\dot{W} = (\dot{W})_D$, then extra heat has to be bought in. Thus in terms of the three options, in decreasing load on the plant operating line: for $[(\dot{W})^*, (\dot{Q}_U)^*]$ both electricity and heat may be sold; for $(\dot{W}, \dot{Q}_U)$ either electricity may be sold, or heat must be bought in.

This has been a general discussion of "off-design" performance of CHP plant. The work by Flink and El-Masri[9] reported in section 4.4.3, also takes account of off-design operation on overall economics, as does that of Treleven et al.[18] who conclude that the important parameters are λ, η_a and η_{CG}.

4.5.3   Meeting Variations in Demand by Cycle Change

In practice a CHP scheme will almost certainly have to respond to changes in demand—the locus of the point $(\dot{W}_D, (\dot{Q}_U)_D)$ on the $(\dot{W}, \dot{Q}_U)$ chart will move from day to night, and from summer to winter.

The obvious internal adaptation of the plant is to run it at other points on its part-load characteristic (supplemented by purchase or sale of heat

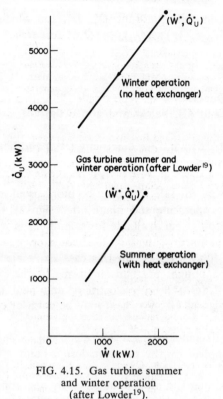

FIG. 4.15. Gas turbine summer
and winter operation
(after Lowder[19]).

and/or electricity). Alternatively, the plant cycle itself may be changed to give a different operating characteristic. One example of such an alteration is that proposed by Lowder,[19] in which for a gas-turbine plant designed with a heat exchanger, the latter may be used when heat demand is low (summer) but not used when heat demand is high (winter), as described in section 3.8.1. The effect of such a change is illustrated in Fig. 4.15, for a gas turbine varying from full (electrical) load to 40% load, and for full recuperation (60% thermal ratio) and no recuperation.

4.6 Summary

This discussion of CHP economics illustrates the considerable complexity of such studies, and how much they are dependent upon initial assumptions. Rather than attempt to draw broad conclusions here, we discuss practical schemes in the next chapter, and defer drawing conclusions until the following chapter, Chapter 6.

References

1. *Combined Heat and Electrical Generation in the United Kingdom Department of Energy Paper No. 35*. HMSO, London, 1979.
2. Williams, R. H. Industrial Cogeneration. *Ann. Review Energy*, **3**, 313–356, 1978.
3. Belding, J. A. *Cogeneration*. MIT Press, Cambridge, Mass., 1982.
4. Kehlhofer, R. Comparison of Power Plants for Cogeneration of Heat and Electricity, *Brown Boveri Review*, 8–80, 504–511, 1980.
5. Lilley, P. Economic Analysis of Industrial CHP Schemes. GEC Power Engineering Report, 1980.
6. Jody, B. J., Daniels, E. J. and Bowman, R. M. *Economics of Industrial Cogeneration*. Proc. Inter Society Energy Conversion Engineering Conference, Vol. I, 534–539, ASME, 1981.
7. Daudet, H. C. and Trimble, S. W. Evaluation Method for Closed Cycle Gas Turbines in Cogeneration Applications. ASME paper 80-GT-176, 1980.
8. Porter, R. W. and Mastanaiah, K. Thermal-Economic Analysis of Heat-Matched Industrial Cogeneration Systems. *Energy*, **7**, 2, 171–187, 1982.
9. Flint, B. B. and El-Masri, M. A. Factors Affecting the Economics of Small, Free-Standing Cogeneration Systems. ASME Paper 84-GT-171, 1984.
10. Marchand, M., Proost, S. and Wilberz, E. A Model of District Heating Using a CHP Plant. *Energy Economics*, 247–257, October, 1983.
11. District Heating Combined with Electricity Generation in the United Kingdom. Department of Energy, Paper No. 20. HMSO, London, 1977.
12. Horlock, J. H. and Owen, R. G. Thermodynamics and Economics of District Heating Using Combined Heat and Power Plant. British Association Paper A21, 1976.
13. Baumann, K. Some Considerations Affecting the Future Development of the Steam Cycle. *Proc. Inst. Mech. Engrs.*, **155**, 125, 1946.
14. Haywood, R. W. *Analysis of Engineering Cycles* (3rd Edition). Pergamon Press, Oxford, 1980.
15. Macadam, J. A., Jebson, D. A., Owen, R. G. and Brogan, R. J. District Heating Combined with Electricity Generation: A Study of Some of the Factors Which Influence Cost-Effectiveness. Department of Energy Report, London, 1981.
16. Cassels, J. M. Some Remarks About District Heating Network Costs. University of Liverpool, Department of Physics Report, January 1980.
17. Heat Loads in British Cities. Department of Energy Paper No. 34. HMSO, London, 1979.
18. Treleven, K., Baugn, J. W. and McKillop, A. A. A Model for the Economic Analysis of Cogeneration Plants with Fixed and Variable Output. *Energy*, **8**, 7, 547–552, 1983.
19. Lowder, J. R. A. Aspects of Matching Complex Industrial Energy Demand Patterns Using Recuperative Gas Turbine Future Energy Concepts, Proc. I.E.E. Conference, London, 1979.

CHAPTER 5

Some Practical CHP Schemes

5.1 Introduction

As has been indicated earlier in this volume, combined heat and power is not new; cogeneration schemes have been operating for many years in several countries. In this chapter, six schemes have been selected for detailed description; the thermodynamics of each is discussed. In some of the examples details of the economics are also included, but in general this commercial information is not available. The six involve the use of the following power plants: back pressure steam turbine; extraction steam turbine; two gas turbines with heat recovery steam boilers (with and without a "bottoming" steam turbine); Diesel engine with a heat recovery steam boiler; and a gas engine with exhaust heating of a water system. They thus cover a wide range of applications and include both CHP/DH (district heating) and CHP/IND (industrial heating).

5.2 Back Pressure Steam Turbine Plant for Aubrugg, Zurich

A back pressure steam plant for CHP/DH has been engineered by Sulzer Brothers for the Aubrugg heat and power station north of Zurich in Switzerland (see Bitterli *et al.*[1]). Two low-pressure (heat only) boilers (117 MW heat)) and three back pressure CHP turbines, each of 45 MW, can supply a total heat load of 465 MW to four main districts (Opfikon (58 MW), Wallisellen (58 MW), Oerlikon (116 MW), Zurich (up to 285 MW)). The heat/power ratio of the CHP plant is

$$\lambda_{CG} = \frac{(465 - 117)}{135}$$

$$= \frac{348}{135}$$

$$= 2.6.$$

Advantage is taken of the requirements of consumers for varying hot water temperatures; the maximum return temperature is 70°C and the maximum supply temperature is 130°C. The latter temperature follows from an

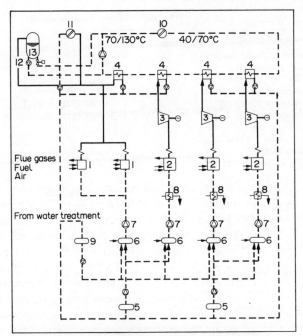

1 Two low-pressure boilers (100 t/h each, 20 bar, 245 °C)
2 Three medium-pressure boilers (250 t/h each, 116 bar, 525°C)
3 Three back-pressure turbosets (about 45 MW each)
4 Steam calorifiers (total thermal output about 465 MW)
5 Two condensate tanks (50 m³, 110°C)
6 Four feedwater tank/deaerators (80 m³, 140°C)
7 Feedwater pumps
8 Feedwater preheaters (140°C to 210°C)
9 Treated water tank (80 m³, 20°C)
10 High temperature hot water for heat consumers (university, hospitals, dwellings)
11 Low-pressure steam for heat consumers (interconnecting line to refuse incinerating plant)
12 Infeed/withdrawal unit
13 Pressure maintenance/expansion vessel

FIG. 5.1. Back pressure steam turbine plant for
Aubrugg (after Bitterli, Leimer and Ruzek[1]).

optimisation calculation (similar to that described in Section 4.4.4.1). With a fixed return temperature, as supply temperature is raised so electrical power output is reduced, the water flow rate is less (pipe work becomes smaller) but heat losses increase. A balance is struck to give minimum price for the heat supply which was the economic criterion used by Bitterli *et al*.

Details of the plant are given in Fig. 5.1 and Table 5.1. The figure shows

TABLE 5.1. *Aubrugg CHP Station—Operating Conditions*

Sulzer back pressure steam turbines (3)	
Electrical power output	135 MW (45 MW each)
Entry pressure	116–120 bar
Entry temperature	525°C
Exhaust (back) pressure	2 bar (or lower)
Exhaust temperature	230°C (or lower)
Heat output	348 MW
Boilers (3 for turbine supply)	
Steam supply	250 t/h each
Feed water temperature	210°C
Boilers (2 "heat-only")	
Steam supply	100 t/h each
Steam pressure	20 bar
Steam temperature	245°C
Heat output	117 MW
Total thermal output	465 MW
CHP plant heat/power ratio	$\dfrac{348}{135} = 2.6$
Overall heat/power ratio	$\dfrac{465}{135} = 3.4$

the three 45 MW back pressure turbines each with one stage of direct contact feed heating and one stage of surface feed heating. They supply exhaust steam to three calorifiers which heat the district water supply from 70°C to 130°C at maximum load. In parallel two "heat-only" steam boilers of high efficiency (92%) supply steam to a calorifier in which the district water supply is also heated. These boilers help to cover the winter demand peaks, while the CHP plant meets the base load.

The use of the two heat-only boilers also gives flexibility in part-load operation. The heat load can drop to 15% of maximum load, and under this condition the return/supply district water temperatures drop to 40°C/70°C respectively.

Bitterli *et al.* report on the economic studies which led to the optimisation of this CHP plant. They emphasise the importance of multi-stage water heating (increasing the electrical power output, i.e. reducing the Z-factor) and utilise a refuse incinerating plant in parallel with the CHP plant to supply the highest temperature requirements with steam. Figure 5.2 shows the energy flow diagram for the scheme. Bitterli *et al.*[1] describe the build-up of the heat load in stages, starting with the heat-only boilers. Their criterion of economic performance is the price of heat (see section 4.2.2) and they show how this price came down as the plant built up to maximum load.

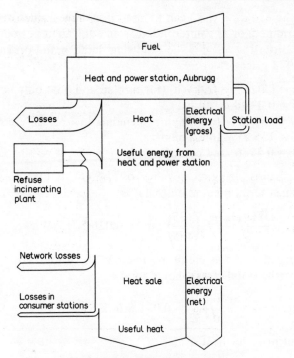

FIG. 5.2. Energy flow diagram for Aubrugg plant
(after Bitterli, Leimer and Ruzek[1]).

Referring to section 4.4.2, the unit price of heat may be expressed in the form

$$(Y_H)_{CG} = \frac{(P_H)_{CG}}{\dot{Q}_U H} = \frac{\beta C_{CG}}{\dot{Q}_U H} + \frac{\dot{F}_{CG}\mathscr{S}_{CG}}{\dot{Q}_U} + \frac{(OM)_{CG}}{\dot{Q}_U H} - \frac{\dot{W}(Y_E)_{BB}}{\dot{Q}_U}, \quad (5.1)$$

where $(P_H)_{CG}$ is the value of heat (Swiss francs p.a.);

   $\dot{Q}_U$ is the heat load (kW);

   $\dot{F}_{CG}$ is the total energy supply to the CHP plant (kW); (i.e. to turbine and heat only boilers);

   C_{CG} is the capital cost of the plant (Swiss francs);

   $\mathscr{S}_{CG}$ is fuel price (S.fr./kWh);

   $(OM)_{CG}$ is the operating and maintenance cost of the plant (S.fr. p.a.);

   $(Y_E)_{BB}$ is the sale price of electricity (S.fr./kWh);

   H is the number of hours operation per year (h p.a.).

This expression is simplified, since load is variable; $\dot{W}$, $\dot{Q}_U$ and $\dot{F}_{CG}$ are variable in practice and have to be integrated over the operating period.

However, the equation (5.1) can be used if average values over a period H are determined, or H is interpreted as an equivalent period of full load operation. Bitterli *et al.* give the following data for full operation of the plant (at 1978 prices):

$$\mathscr{S}_{CG} = 0.0215 \text{ (S.fr./kWh) (for turbine and heat only boilers)};$$
$$(Y_E)_{BB} = 0.09 \text{ (S.fr./kWh)};$$
$$\dot{Q}_U = 465 \text{ MW (maximum) (including low-pressure boilers)};$$
$$\dot{W} = 135 \text{ MW (maximum)};$$

and $\beta = 0.5786$ may be deduced for $i = 4\%$, $N = 30$ years).

With these figures (using mean values of $\dot{F}_{CG}$, $\dot{Q}_U$, $\dot{W}$) the credit for electricity (the last term in equation (5.1)) is

$$\frac{\dot{W}(Y_E)_{BB}}{\dot{Q}_U} = \left(\frac{135}{465}\right)0.09 = 0.026 \text{ S.fr./kWh}.$$

Bitterli suggests that this credit is some 63.2% of total costs so that the first three terms (total costs) should total

$$\frac{0.026}{0.632} = 0.041 \text{ S.fr./kWh,}$$

giving a unit price for heat of

$$(Y_H)_{CG} = 0.041 - 0.026 = 0.015 \text{ S.fr./kWh.}$$

Bitterli also suggests that the three terms making up the total costs should be

$$\frac{\beta C_{CG}}{\dot{Q}_U H} = 0.203 \cdot 0.041 = 0.0084 \text{ S.fr./kWh,}$$

$$\frac{\dot{F}_{CG}\mathscr{S}_{CG}}{\dot{Q}_U} = 0.660 \cdot 0.041 = 0.0272 \text{ S.fr./kWh,}$$

$$\frac{(OM)}{\dot{Q}_U H} = 0.137 \cdot 0.041 = 0.0056 \text{ S.fr./kWh.}$$

(Direct calculation of the second term, with $\dot{F}_{CG} = 135 + 465 = 600$ MW, $\dot{Q}_U = 465$ MW, $\mathscr{S}_{CG} = 0.0215$ S.fr./kWh, gives

$$\frac{\dot{F}_{CG}\mathscr{S}_{CG}}{\dot{Q}_U} = \left(\frac{600}{465}\right)0.0215 = 0.0278 \text{ S.fr./kWh,}$$

which suggests that these estimates of costs and heat prices are reasonable.)

The dominant factor in total costs is the cost of fuel (66%). Amortisation and interest costs (capital charges) are surprisingly small (20.3%)

but the discount rate of 4% is low, and the plant life assumed is long ($N = 30$ years). Utilisation of an equivalent 2,600 hours per annum at full load (a figure derived by dividing the total heat output by the maximum heat load of 465 MW, Bitterli, private communication[2]) would suggest a capital cost of

$$C_{CG} = \frac{0.0084 \cdot 2.6 \cdot 10^3 \cdot 465 \cdot 10^3}{0.05786}$$

$$= 1.77 \cdot 10^8 \text{ S.fr.,}$$

and

$$\frac{C_{CG}}{\dot{W}} = \frac{1.77 \cdot 10^8}{1.35 \cdot 10^3} = 1.3 \cdot 10^3 \text{ S.fr./kW.}$$

This is comparable with the capital cost of back pressure turbines used in the Marshall report[3] and given in section 4.4.3. However, it should be noted that the value of C_{CG} derived above is a "mean" capital cost covering both the back-pressure turbine and the heat only boiler plant.

5.3 Extraction Steam Turbine Plant in USSR

A large scale extraction steam turbine which has been in operation in the USSR since 1972 is described by Oliker[4] on the basis of several Russian publications.

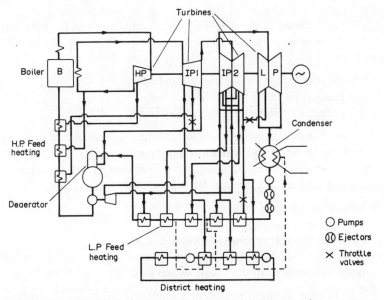

FIG. 5.3. 250 MW CHP steam turbine (after Oliker[4]).

TABLE 5.2. *USSR 250 MWh Extraction Steam Turbine—Some Operating Conditions*

Extraction Steam Turbine	(4 cylinders, HP, IP1, IP2, LP)
Throttle flow	$2.1 \cdot 10^6$ lb/h ($0.95 \cdot 10^3$ t/h)
Absolute steam pressure (entry)	3400 psia (24.1 MPa)
Steam temperatures (entry and reheat)	1040°F, 1049°F (560°C, 565°C)
Absolute steam pressures (entry to IP2)	78.2 psi (539.2 kPa) ⎱ in condensing
(entry to LP)	11.4 psi (78.6 kPa) ⎰ mode
De-aerator pressure	85 psi (586.1 kPa)
District heating extraction pressures	
stage (iii) (along I.P.)	8.5–38.5 psi (58.6–265.4 kPa)
stage (iv) (exhaust from I.P. entry to L.P.)	7.2–21.3 psi (49.6–146.9 kPa)
Electric power output	250 MW
Heat load	385 MW
Z factor (Lost work/Heat load)	0.13
Heat/power ratio	1.54
Percentage throttle flow used for district heating	70–80%

The steam turbine delivers 250 MW of electrical power and 385 MW of heat. The plant is shown diagrammatically in Fig. 5.3. The maximum capacity of the steam turbine with the district heat disconnected is 300 MW, suggesting a Z factor (lost work/heat utilized) of 50/385 = 0.13, which is (desirably but surprisingly) low. However, the plant is sophisticated, with many extraction points for both feed heating and district heating, and this could enable a low value of Z to become possible.

Table 5.2 gives some of the major operating conditions of the plant (not all are available in ref. 4). Reheat is supplied between the HP and IP1 cylinders. Three high pressure feed heaters are fed from

(i) an extraction point along the HP cylinder;
(ii) the exhaust from the HP cylinder; and
(iii) an extraction point along the IP1 cylinder.

A de-aerator is supplied with steam bled from the exhaust of IP1. The boiler feed pump is turbine driven with steam extracted from the IP1 cylinder and returning to the main turbine upstream of the IP2 cylinder. Five low pressure feed heaters are fed with steam

(i) from this returning stream;
(ii) and (iii) from two extraction points along the IP2 cylinder;
(iv) from the exhaust of the IP2 cylinder; and
(v) from an extraction point along the LP cylinder (the last heater also takes in steam from the gland system).

The district heating water is heated first in a condenser also fed by steam from the last source (v) above (i.e. in a gland steam condenser). It is then heated in two heat exchangers supplied with steam from the extraction

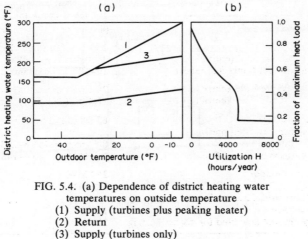

FIG. 5.4. (a) Dependence of district heating water
temperatures on outside temperature
(1) Supply (turbines plus peaking heater)
(2) Return
(3) Supply (turbines only)
(b) Heat load as a function of utilisation (after
Oliker[4]).

point at exhaust from IP2 (source (iv) above) and from the lower extraction point along the IP2 cylinder (source (iii) above). Finally the district
heating water may be heated in a "peaking" water heater, for maximum
heat load conditions.

Water temperatures for the district heating (and the extraction pressures
from the turbines) are closely related to outdoor temperatures, and hence
to the overall heat load, as indicated in Fig. 5.4(a). High supply temperatures required in cold weather are achieved without raising the turbine
extraction temperature (and pressure) to too high a level (and hence
reducing electrical power output) by use of the "peaking" water heater at
the end of the district water heating chain. The two extraction pressures
(absolute) range from 8.5 to 38.5 psi (58.6 to 265.4 kPa) (upper) and 7.2
to 21.3 psi (49.6 to 146.9 kPa) (lower) as the outdoor temperature varies
over the range shown in Fig. 5.4(a). The optimum distribution of the heat
load between the heaters is a complex function of many factors (the overall
heat load, the outdoor temperatures, etc.). Figure 5.4(b) shows a typical
heat load variation over a year's operation (similar curves apply to other
plant, such as the back pressure plant described in section 5.2). Oliker does
not give details of the economics of this 250 MW plant, but it is clearly one
of the most sophisticated and complex CHP plants in existence.

5.4   Gas Turbine Plant (with Heat Recovery Steam Generator)
for Beilen (Netherlands)

A gas turbine CHP scheme, with a heat recovery steam generator
(HRSG), has been installed at the DOMO plant in Beilen in the Nether-

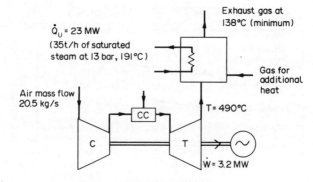

Ruston TB gas turbine

FIG. 5.5. CHP plant for Beilen (HRSG fired,
$(\lambda_{CG})_2 = 23/3.2 = 7.19$).

lands.[5] The plant, which produces dairy products, originally took its electric power (up to 3.2 MW) from the grid and its heat load was met by two gas-fired boilers with a steam production of 25 t/h at 13 bar.

The CHP plant which replaced these two separate energy supplies is shown in Fig. 5.5. A Ruston TB gas turbine (rated at 3.65 MW) can meet the electrical demand of 3.2 MW, and is connected to the grid so that excess electrical power can be sold. (Standby electricity is also available in the event of plant breakdown or maintenance.) The gases leaving the turbine exhaust into a heat recovery steam generator, and at full gas turbine power 12 t/h of saturated steam at 191°C can be produced at 13 bar. Five supplementary gas burners can be engaged (the gases leaving the turbine contain 16% oxygen) to increase the steam production to 35 t/h. Gases leave the exhaust stack at 138°C under maximum load conditions.

The flow parameters for the two plant operating conditions—of HRSG unfired and HRSG fired—are given in Table 5.3. For the first operating condition (HRSG unfired) the heat load is estimated at 7.5 MW. Similarly, for the second condition (HRSG fired), when 35 t/h saturated steam is raised then the heat load is 23 MW. The values of heat to power ratio are

$$\left(\frac{7.5}{3.2}\right) = 2.34, \quad \text{and} \quad \left(\frac{23}{3.2}\right) = 7.19 \text{ respectively.}$$

The conversion of the DOMO plant to CHP provides a good example of the "change" type of analysis—from separate heat and electricity supplies to a "matched" gas turbine plant (section 3.3); this is the case when Beilen operates with no export of electricity.

TABLE 5.3. *Ruston TB Gas Turbine* (Rating 3.65 MW)

Alternator power output	3.2 MW
Air mass flow rate	20.45 kg/s
Pressure ratio	7:1
Maximum temperature	890°C
Thermal efficiency	0.23

Heat Recovery Steam Generator (HRSG)

Unfired	Steam (saturated) mass flow rate	12 t/h
	Steam pressure	13 bar
Fired	Steam (saturated) mass flow rate	35 t/h
	Steam pressure	13 bar

Performance parameter

$$\psi = \frac{\text{(Heat transferred to steam)}}{\text{(Fuel energy supplied)}} = 1.34$$

Figure 5.6 shows diagrammatically the original plant with unit work output, as in section 3.2. The *rate* of energy consumption is

$$\dot{F}_{REF} = \frac{1}{\eta_C} + \frac{\dot{\lambda}}{\eta_B}, \tag{5.2}$$

where η_C is the efficiency of the "grid" electrical power plant and η_B the efficiency of the heat only boiler.

For operation of the CHP plant in the first operating condition with the HRSG unfired, the energy consumption is simply

$$\dot{F}_1 = \frac{1}{\eta_{CG}}, \tag{5.3}$$

where η_{CG} is the thermal efficiency of the cogeneration plant, and the energy utilisation factor and fuel energy savings ratio are

$$(EUF)_1 = \frac{[1 + (\dot{\lambda}_{CG})_1]}{\dot{F}_1}$$

$$= \eta_{CG}[1 + (\dot{\lambda}_{CG})_1], \tag{5.4}$$

$$(FESR)_1 = \frac{\dot{F}_{REF} - \dot{F}_1}{\dot{F}_{REF}}$$

$$= 1 - (\eta_C\eta_B)/\{\eta_{CG}[\eta_B + (\dot{\lambda}_{CG})_1\eta_C]\}. \tag{5.5}$$

respectively (see equations (3.7) and (3.9)).

Original plant

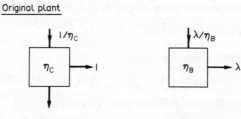

$$F_{ref} = \frac{1}{\eta_C} + \frac{\lambda}{\eta_B}$$

HRSG Unfired

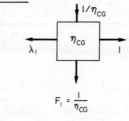

$$F_1 = \frac{1}{\eta_{CG}}$$

HRSG Fired

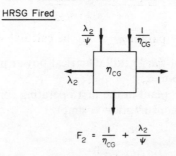

$$F_2 = \frac{1}{\eta_{CG}} + \frac{\lambda_2}{\psi}$$

FIG. 5.6. Plant fuel consumption.

For operation of the CHP plant in the second operating condition, with the HRSG fired, the energy used is

$$\dot{F}_2 = \frac{1}{\eta_{CG}} + \frac{(\dot{\lambda}_{CG})_2}{\psi}, \qquad (5.6)$$

where ψ is a performance parameter for the HRSG defined as

$$\psi = \frac{\text{Heat transferred (to steam)}}{\text{Fuel Energy Supplied}}$$

(Discussion of the performance of a heat recovery steam generator leading to the definition of ψ (which may be greater than unity) was given in section 3.4.3.)

The fuel energy savings are thus

$$\dot{F}_{REF} - \dot{F}_2 = \left[\frac{1}{\eta_C} + \frac{(\dot{\lambda}_{CG})_2}{\eta_B}\right] - \left[\frac{1}{\eta_{CG}} + \frac{(\dot{\lambda}_{CG})_2}{\psi}\right], \tag{5.7}$$

and the fuel energy savings ratio is

$$(FESR)_2 = \frac{\dot{F}_{REF} - \dot{F}_2}{\dot{F}_{REF}} = 1 - \frac{\eta_C \eta_B[\psi + \eta_{CG}(\dot{\lambda}_{CG})_2]}{[\eta_B + (\dot{\lambda}_{CG})_2 \eta_C]\eta_{CG}\psi}. \tag{5.8}$$

We may note that there are no fuel savings unless

$$\frac{1}{\psi} < \frac{1}{\eta_B} - \frac{1}{(\dot{\lambda}_{CG})_2}\left(\frac{1}{\eta_{CG}} - \frac{1}{\eta_C}\right).$$

The energy utilisation factor is

$$(EUF)_2 = \frac{1 + (\dot{\lambda}_{CG})_2}{\dot{F}_2}$$

$$= \frac{[1 + (\dot{\lambda}_{CG})_2]\eta_{CG}\psi}{\psi + (\dot{\lambda}_{CG})_2 \eta_{CG}}. \tag{5.9}$$

For the DOMO plant, these parameters may be calculated as follows.

With HRSG unfired $(\dot{\lambda}_{CG})_1 = 2.34,$

$\eta_{CG} = 0.23;$

and assuming $\eta_C = 0.4,$

$\eta_B = 0.9;$

$$(FESR)_1 = 1 - \frac{0.36}{0.23(0.9 + 2.34 \cdot 0.4)} = 0.143,$$

$$(EUF)_1 = 0.23(3.34) = 0.77.$$

With HRSG fired, $(\dot{\lambda}_{CG})_2 = 7.19,$

$\eta_{CG} = 0.23;$

and again assuming $\eta_C = 0.4,$

$\eta_B = 0.9;$

$$(FESR)_2 = 1 - \frac{0.36(1.34 + 0.23 \cdot 7.19)}{(0.9 + 0.4 \cdot 7.19)0.23 \cdot 1.34} = 0.075$$

$$(EUF)_2 = \frac{(1 + 7.19)0.23 \cdot 1.34}{(1.34 + 0.23 \cdot 7.19)} = 0.85.$$

Total running time per year of the DOMO plant is very high—of the order of 7500–8000 hours per year—but this includes part-load operation. Details of fuel and electricity costs are not available, so the economic analysis of Chapter 4 cannot be applied directly. However, savings of 10^6 Dfl. per year are estimated on fuel and electricity costs resulting in a pay-back period of less than 6 years.

5.5   Gas Turbine Plant (with Heat Recovery Steam Generator and Steam Turbine) for Saarbrücken

A more complex combined plant (a gas turbine with a heat recovery steam generator (HRSG) and steam turbine) was installed in 1974 in the city of Saarbrücken and has been described by Flad.[6] Part of the output from the HRSG is used to supply steam to the steam turbine, which both generates additional electrical power and meets a district heating load. Supplementary firing of the HRSG increases the steam supply and the additional power output. The plant is referred to as "STAG" (steam and gas turbine).

The full operation of the plant is illustrated diagrammatically in Fig. 5.7. The gas turbine exhausts into the HRSG and thence to the exhaust stack. Steam generation in the HRSG is 47 t/h at 430°C under normal operation, but may be increased to 85 t/h at 510°C by supplementary firing. The steam is expanded in a steam turbine, but some steam may be extracted for supplying heat through heat exchangers to a district heating water supply. With the heat load on, district heating water first passes through

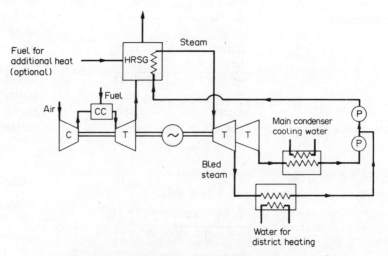

FIG. 5.7. Combined CHP plant for Saarbrücken (after Flad[6]).

the exhaust gas cooler (picking up some 5°C and dropping the exhaust gases to 110°C) if the gas turbine is fuelled with natural gas. If distillate fuel is used in the gas turbine this extra economiser is by-passed, to avoid dew-point corrosion. The district heating water is then heated by extracted steam condensing in a heat exchanger.

There are essentially four modes of operation of the "bottoming" steam generator and turbine plant.

(a) HRSG unfired, no steam extraction;
(b) HRSG unfired, steam extraction;
(c) HRSG fired, small steam extraction;
(d) HRSG fired, large steam extraction.

Details of these four plant operating conditions are given in Table 5.4. There is thus considerable flexibility in this plant, with the ratio of heat load to electrical load (λ_{CG}) ranging from zero (case (a)) up to 1.3 (case (b)), for the unfired HRSG; and from 0.44 (case (c)) up to 1.3 (case (d)), for the fired HRSG.

Flad does not give details of the economic performance of the Saarbrücken plant, but since the STAG plant has supplied approximately 20% of the electrical power required by the city, it would be logical to use the analysis of section 4.2.1 to calculate the price of electricity supplied and compare it with standard plant.

The Saarbrücken scheme obtains both power output *and* district heating (CHP/DH) from the steam generated in the HRSG. Allen and Kovacik[7] describe several other gas turbine schemes used for industrial combined heat and power (CHP/IND). They discuss cases where steam from the HRSG may be used as process steam (as in the Beilen plant of section 5.4) and/or power generation in a steam turbine (as in the Saarbrücken plant).

Steam may also be recirculated (fully or partially) for injection in the gas turbine combustion chambers to give NO_x control and extra power. The performance of a plant operating in this way is described by Messerlie and Strother.[8] The technique can be used in a cogeneration plant; variation of the amount of steam abstracted for recirculation controls the heat to power ratio (λ_{CG}).

5.6 Diesel Engine Plant (with Heat Recovery Steam Generator) for Hereford

The Midlands Electricity Board (UK) has developed a complex CHP plant based on Diesel generation, supplying heat to two industrial consumers and power to the grid. Figure 5.8 is a simplified diagram of the plant (showing one of the two engines), taken from ref. 9).

Electrical power is obtained from a generator driven through a gear-

TABLE 5.4. *Saarbrücken CHP Plant*

Gas Turbine   (MS-5001 Alsthom/General Electric)

Mass flow	410 t/h
Output (electrical)	23.5 MW
Fuel energy supplied	87.7 MW
Basic efficiency	26.8%
Fuel	Natural gas/Light distillate

Steam Turbine

		λ_{CG}	(EUF)
With unfired HRSG (no fuel energy supplied to HRSG)			
(a)	Steam mass flow		
	47 t/h at 430°C		
	Power output (no extraction)		
	9.25 MW		
	Heat production	$\dfrac{5.82}{32.75} = 0.18$	$\dfrac{38.57}{87.7} = 0.44$
	5.82 MW		
(b)	Steam mass flow		
	47 t/h at 430°C		
	Power output (steam extraction)		
	4.6 MW		
	Heat production (steam extraction)	$\dfrac{42.2}{28.1} = 1.5$	$\dfrac{70.3}{87.7} = 0.80$
	42.2 MW		
With fired HRSG (35.2 MW fuel energy supplied to HRSG—natural gas)			
(c)	Steam mass flow		
	85 t/h at 510°C		
	Power output (25 t/h steam extraction)		
	20.05 MW		
	Heat production	$\dfrac{23.2}{43.55} = 0.53$	$\dfrac{66.75}{87.7 + 35.2} = 0.54$
	23.2 MW		
(d)	Steam mass flow		
	85 t/h at 510°C		
	Power output (60 t/h steam extraction)		
	13.8 MW		
	Heat production	$\dfrac{52.3}{37.3} = 1.4$	$\dfrac{89.6}{87.7 + 35.2} = 0.73$
	52.3 MW		
	Extraction pressures		
	0.7–2.2 bar		

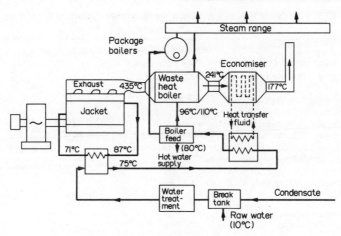

FIG. 5.8. CHP plant for Hereford (after Shepherd[9]).

box by a Diesel engine (10,400 H.P. 16 cylinder vee type medium speed), supplying 7.48 MW. Exhaust gases from the engine pass first through a heat recovery steam generator (unfired) and then an economiser. The exhaust gases leave the engine at a mean temperature of 815°F (436°C) and drop to 466°F (242°C) across the HRSG. A further drop in temperature to 350°F (177°C) occurs across the economiser.

The plant mainly provides process steam, with very little condensate returned from the consumers. Figure 5.8 shows the main water flow entering at 50°F (10°C) and, after treatment, first being heated to 167°F (75°C) in a heat exchanger, by jacket cooling water from the engine. This flow is joined by returned condensate, and part of the resulting water flow is heated to 176°F (80°C) in the economiser and supplies a small "district heating" load. The major part of the water flow is heated to a higher temperature (187°F/230°F or 86°C/110°C) in the economiser, and pumped to the HRSG or stand-by oil-fired boilers, which supply the process steam. Since little condensate is returned the water flow is essentially "once-through".

The operating conditions of the Hereford CHP plant are given in Table 5.5. At maximum load, a power output of 14.96 MW and a heat output of 13.4 MW have been achieved.

The plant illustrates that a high energy utilisation factor (0.75) can be obtained, but at a relatively low heat to power ratio ($\lambda_{CG} = 0.895$). This confirms the work of Porter and Mastanaiah referred to in Chapter 3; since the thermal efficiency of the Diesel engine is already high, the available heat for "useful rejection" is less than in a plant of lower efficiency,

TABLE 5.5. *Diesel Engine CHP Plant for Hereford—Operating Conditions at Design Point (Thermodynamic Performance)*

Diesel Engines	$2 \times 10{,}400$ HP
Electrical Power Output	7.48 MW each
Thermal efficiency (maximum)	39.6%
Heat Output per engine	6.7 MW

Heat rejection from jacket cooling water	2.7 MW	low grade heat
Heat rejection from exhaust gases to economiser	1.0 MW	
Heat rejection from exhaust gases to heat recovery		high grade heat
steam generators	3.0 MW	

Steam supply (saturated) 4500 kg/h at 21 bar

Heat/power ratio $\lambda_{CG} = \dfrac{6.7}{7.48} = 0.895$

Energy utilisation (of CHP plant alone)

$$(EUF)_{CG} = \eta_{CG}(\lambda_{CG} + 1)$$
$$= 0.396\,(1.895)$$
$$= 0.75$$

such as a back-pressure turbine. However, the heat demand at Hereford may rise to a maximum of 40 MW, so four oil-fired boilers (using the same fuel as the prime-movers) were installed to meet the extra heat load.

The detailed economics required to price the heat are not available. But since the electricity is generated for the utility itself, it would be logical to use the "change" type of analysis adopted in the Marshall report[3] in which it was assumed that the overall production of electricity by the utility had to be maintained, and the electricity price to consumers not changed. However, the efficiency of power generation by the Diesel engines may be different from the efficiency of power generation by the grid station they are effectively replacing, and if there is an associated penalty on fuel consumption it has to be subtracted from the fuel saving resulting from the reduction in "heat-only" boiler capacity following introduction of the CHP plant. (This approach was given in section 4.4.2(b), for a back pressure plant.) With the overall money savings on fuel estimated (less any increase of maintenance costs) these annual savings have to be balanced against increased capital costs. The form of calculating $(NPV)_b$ given in equation (4.43) is a convenient way of completing this "change" type of analysis, which is valid when the electricity utility itself is introducing the CHP scheme, and selling the heat to an industrial customer.

5.7   Gas Engine Plant (with Heat Recovery Water Heater) for the Open University

Several CHP options for meeting the electricity and heat demands of the UK Open University have been considered, including gas turbine, back pressure steam turbine, and Diesel engine schemes (ref. 10). In all three of these options, since the size of the prime mover would be small the plant would be relatively inefficient (i.e. of low thermal efficiency).

A fourth scheme now favoured (but not yet implemented) uses a turbo-charged spark-ignition gas engine, with heat rejected (from both the exhaust gas stream and the jacket water cooling system) to circulating water for space and domestic hot water heating (Fig. 5.9). The gas engine would run on the existing natural gas supplied to the site but this gas would have to be compressed in a gas compressor requiring extra power.

The heat and electricity demands for the site are similar to those shown in Fig. 5.9. The maximum electrical demand (winter) is 2.1 MW at a power factor of 0.95; but this may drop in summer months to about 1.7 MW. The lowest electrical demand is 0.85 MW (winter) and 0.7 MW (summer).

Table 5.6 gives the operating conditions of the CHP plant. Under the scheme proposed, the gas engine produced 700 kW (electrical), together with 460 kW (thermal), provided at a supply water temperature of 85°C with a return water temperature of 71°C.

Analysis of the economics of this CHP scheme was carried out in a conventional manner, by comparing the costs of the new scheme with those of the existing plant (bought-in electricity and gas consumption). Study of the proposal provides an illustration of how plant operation varies over each day, and round the year. Savings must be estimated by summing the

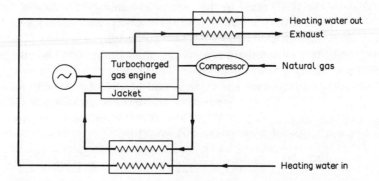

FIG. 5.9. CHP plant for Open University.

TABLE 5.6. *Gas Engine CHP Plant for Open University—Operating Conditions*

Electrical power output	700 kW
Heat output	
Exhaust gas heat exchanger	460 kW
Jacket water (if used)	160 kW
Gas consumption	6900 ft³/h (195.4 m³/h)
Efficiency	36%
Water return temperature	71°C
Water supply temperature	
(heat exchanger only)	85°C

separate contributions over several periods. In this case three periods were identified as follows.

(a) CHP plant operating at full output (electricity and heat) for 36 winter weeks during the daytime (07.30–19.30), when the total space and domestic heat load exceeds 460 kW; extra electricity and gas for heat-only boilers are bought in.

(b) CHP plant operating weekday evenings (19.30–00.30) during the 36 winter weeks, and day and evening period (07.30–00.30) at weekends during those weeks; extra electricity and gas for heat-only boilers are bought in.

(c) CHP plant operating for the remainder of the year during day and evening periods (07.30–00.30) to generate electricity and supply domestic hot water only (replacing immersion heaters) i.e. for 10% of the total heat load; extra electricity only is bought in if required.

As an example of the calculation of money savings consider the first period (a). The daytime heat demand is in excess of the basic supply of 460 kW in 36 (five-day) weeks for 11.5 hours each day. The electricity generated per year by the CHP plant in this period (assuming 95% availability) is $(36 \times 5 \times 11.5) \times 0.95 \times 700 = 1.377 \cdot 10^6$ kWh, and this leads to a saving on bought-in electricity of $1.377 \cdot 10^6$ (kWh) $\times 0.0362$ (£/unit) = £49,900 p.a. In addition there is a saving on maximum demand charges of £5,800 p.a. giving a total saving of £55,700 p.a.

The saving on heat over the same period of 2070 hours, with an availability of 95% and heat generation of 460 kW is 0.905×10^6 kWh $(30.9 \cdot 10^3$ therms). The value of this heat (i.e. the savings on gas consumption in a gas boiler of 85% efficiency) would be

$$\frac{30.9 \cdot 10^3}{0.85} \text{ (therms)} \times 0.321 \text{ (£/therm)} = £11,700 \text{ p.a.}$$

The total cost of operating convention plant in periods (a) would therefore have been M_C = £67,400 p.a.

The amount of gas supplied to the cogeneration plant (i.e. to the gas engine of 36% thermal efficiency) is given by

$$\frac{1.377 \cdot 10^6}{0.36} \text{ kWh} = 130.5 \cdot 10^3 \text{ therms,}$$

and its cost is

$$M_{CG} = 130.5 \cdot 10^3 \times 0.321 = \pounds 41,900 \text{ p.a.}$$

Additional maintenance costs are estimated at 7% of fuel costs

i.e. $\Delta(OM) = \pounds 2,900$ p.a.

The operating surplus for these periods (a) over the year is

$$\Delta M_a = M_C + (OM)_C - [M_{CG} + (OM)_{CG}]$$
$$= M_C - M_{CG} - \Delta(OM), \hspace{2cm} (5.10)$$

and for this case is

$$\Delta M_a = (\pounds 67,400 - \pounds 41,900 - \pounds 2,900) \text{ p.a.}$$
$$= \pounds 22,600 \text{ p.a.,}$$

during the period (a) within the year.

Analyses of the other operating periods (b) and (c) give surpluses of $\Delta M_b = \pounds 14,900$ p.a. and $\Delta M_c = \pounds 5,700$ p.a. respectively, and a total surplus of $\Delta M = \Delta M_a + \Delta M_b + \Delta M_c = \pounds 43,200$ p.a.

Capital costs (ΔC) for changing to cogeneration were estimated as follows:

Gas engine and generator	£260,000;
Gas booster	£30,000;
Installation and commissioning	£45,000;
Total =	£335,000

At first sight the proposed CHP plant did not seem economic, since without any DCF type of analysis the simple short payback period (PBP = $\Delta C/\Delta M$) was clearly substantial—some 8 years. Miscellaneous additional costs associated with plant installation had not been included (for example it was assumed that the "extra" heat and electricity supplies remained available without additional stand-by charges, since supply lines had been installed for the original plant).

However, a short pay-back period as demanded by industry was not necessarily required here; capital might be available at low cost, as an incentive to reduce annual running charges. It was concluded therefore that the scheme might become marginally economic, particularly if capital cost (mainly that of the engine) could be reduced, and further study was required.

Using the simple DCF analysis of section 4.3.1.1, the net present value would be

$$(NPV)_{(i)} = \frac{\Delta M}{_N f_{AP}} - \Delta C. \tag{5.11}$$

With a discount rate of 5%, and a plant life of 20 years, $_N f_{AP} = 0.08024$. The break-even capital cost (given by putting (NPV) (i) equal to zero) is therefore

$$(BECC) = \frac{\Delta M}{_N f_{AP}} = \frac{43,200}{0.08024} = £538,385,$$

which would suggest the scheme to be feasible under such financial conditions. However, with a discount rate of 10%, $_N f_{AP} = 0.11746$, the break-even capital cost is less, $(BECC) = £367,785$ which illustrates the sensitivity of break-even capital cost to discount rate.

This study and costing was based on a 750 H.P. engine that would operate at 60% of its continuous full load rating. Further investigation into use of a higher speed engine that would operate closer to its maximum rating has indicated that capital cost might be halved (but with the probability of increased maintenance cost and shorter plant life).

5.8 Discussion

The examples described in this chapter illustrate the great variety of CHP schemes, and the diversity of analyses that may be relevant in each case, particularly for economic assessment. No overall conclusions are drawn from these practical cases except for the general point made at the beginning of this book—that fuel savings are simply determined and are usually substantial, but that economic justification is much more complex and difficult.

References

1. Bitterli, J., Leimer, H. J. and Ruzek, W. Combined District Heat and Power Station, Aubrugg, Zurich. *Sulzer Technical Review*, **3**, 87–92, 1980.
2. Bitterli, J., Leimer, H. J. and Ruzek, W. Private Communication, 1985.
3. Combined Heat and Electrical Generation in the United Kingdom Department of Energy Paper No. 35. HMSO, London, 1979.
4. Oliker, I. Steam Turbines for Cogeneration Power Plants. *Trans. ASME Journal of Engineering for Power*, **102**, 482–485, 1980.
5. Gas Turbine Total Energy Improvers Plant Economy. Ruston Gas Turbine Report. *Diesel and Gas Turbines Worldwide*, October 1980.
6. Flad, J. Operation of a Combined District Heating and Power Production Scheme with a Combined Gas Turbine. Lecture Series. Von Karman Institute for Fluid Dynamics. Rhode Saint Genese, Belgium, 1978.

7. Allen, R. P. and Kovacik, J. M. "Gas Turbine Cogeneration—Principles and Practice" *Trans. ASME Journal of Engineering for Gas Turbines and Power*, **106**, 4, 725–730, 1984.
8. Messerlie, R. L. and Strother, J. R. Integration of the Brayton and Rankine Cycles to Maximise Gas Turbine Performance—A Cogeneration Option. ASME Paper 84-GT-52, 1984.
9. Shepherd, G. T. Combined Heat and Power. Local Power and Heat Generation: A New Opportunity for British Industry. Open University Conference, Interscience Enterprises Ltd, Jersey, 1982.
10. Gray, D. V. Private Communication, 1985.

CHAPTER 6

Summary

6.1 Introduction

It is clear from the preceding chapters that the selection of a CHP plant, incorporating thermodynamic and economic analysis, is a very difficult and complex problem. In general, the technology involved in plant construction and operation is well established. As the Marshall report[1] indicated "there appear to be no technical difficulties in the construction of district heating networks or in developing suitable CHP plant". This is also true of industrial CHP.

Further there is generally no question that replacement of existing plant (for the separate generation of electrical power and heat supply) by CHP plant (for cogeneration) leads to substantial energy saving. The estimate of such saving is a relatively straightforward matter and follows from standard thermodynamic cycle analysis. However, the ratio (λ_D) of the heat load to the power demand (whether based on mean or minimum values) is a critical factor in the choice of the type of plant, and flexibility in "off-design" performance to meet variations in heat and electricity requirements (and their ratio) is also important.

Choice of the right plant from the thermodynamic point of view can improve the economic case for CHP, but even with the best possible choice the overall economic case may not be sufficiently attractive to enable the scheme to go ahead.

6.2 District Heating (CHP/DH)

Economies of scale are important for CHP/DH to be viable. To quote the early report of the Working Party of the Marshall group:[2] "Small CHP schemes are not economically attractive and would give only limited energy savings, since most small CHP plants are relatively poor converters of input fuel into electricity. Large CHP schemes are more attractive because they can use large power stations which give substantial energy savings and produce both heat and electricity more economically". They can produce heat "at the power station fence" at a cost which can be as little as half that from a boiler. It is the cost of transmission and distribution which increases the cost of heat to the consumer and makes even large schemes marginally economic in many cases.

The conclusions in the subsequent main report from the Marshall group (on CHP/DH in the medium term) are worth quoting in full.

"CHP for inner-city and large city developments using medium/large turbines is an attractive economic option in national resource terms compared with other heating alternatives if lower discount rates (e.g. 5%) and higher fuel prices (e.g. doubling by the year 2000) are assumed. CHP becomes less attractive at higher discount rates. Since risk and uncertainty were not accounted for in the calculations, a detailed study of CHP/DH for actual locations will be required to examine those factors dependent upon site;

medium/large CHP turbines (200–660 MW) are more economic and save more energy than small CHP plant although it may be economic to heat small developments from local power stations in particular circumstances;

maximum energy savings will only be achieved when heating networks in cities are connected up to medium-to-large turbine CHP plant, and the timescales for doing this are long. Heat-only boiler plant and/or small CHP plant could assist in the build-up phase."

The Marshall calculations for the longer term were more speculative. The assumptions for CHP were similar (see the tables in Chapter 4) but additional assumptions had to be made about competing systems, and the reader is referred to the report for those assumptions and the results of the calculations. The conclusions of the group for the longer term were as follows:

"Medium/large CHP plant serving areas of high heat load densities in cities looks an attractive economic and energy saving option compared with other developed forms of heating. In the absence of natural gas and oil in the longer term, other heating alternatives such as off-peak electricity or SNG (substitute natural gas) would prevail in a large market outside the areas of high-density heat load."

The critical factors for CHP/DH which emerged from the Marshall studies, and the subsequent report by Macadam et al., were

(i) housing density;
(ii) discount rate;
(iii) fuel-price factor (the rates of escalation of the various fuel costs);

and to a lesser extent

(iv) transmission distance.

Subsequently, in further work[3] the supply and return temperatures of the water for district heating appeared as additional critical parameters. The *distribution* of the heat load in a city was shown to be important, as well as the mean housing density.

The significance of these parameters was illustrated in Chapter 4; in Fig. 4.5 and 4.6 for (i), (ii) and (iii); in Fig. 4.8 for (iv); and in Fig. 4.7 for water temperatures. Figure 4.8 confirms the "rider" to the Marshall conclusion, that "it may be economic to heat small developments from local power stations in particular circumstances".

An interesting summary chart (Fig. 6.1) was provided by Hewitt and Owen[4] to illustrate the two important effects of discount-rate and fuel price factor (the ratio of fuel price in AD 2000 to "present" fuel price (in 1976)). CHP/DH really only becomes the best option at low discount rate and high fuel price factor, for a housing density (D) of 20 dwellings per acre. The boundaries shown in the figure change with the choice of D, but the picture will be broadly of the same shape. However, it should be remembered that the subsequent work reported in ref. 3 concluded that when the distribution of heat load was represented more realistically, the network costs estimated by Marshall were reduced by 50%, leading to a reduction of about 15% in the annual costs of supplying heat.

6.3   Industrial and Other Applications (CHP/IND)

The position on CHP/IND is less clear than that for CHP/DH; it is difficult to generalise, for the appropriate choice of plant varies from one application to another. But overall the case for CHP/IND is stronger.

As far as the thermodynamics is concerned, the selection of type of plant not only depends largely on the heat to power ratio (see Fig. 3.19) but also on the size of the heat and electricity demands. (For example, in the last case considered in Chapter 5 (section 5.7), with small heat and electricity demands each of the order of 1 MW, steam and gas turbines were quickly eliminated because of low efficiency in this range.) The variation in demand, daily and seasonal, is also important, but it should

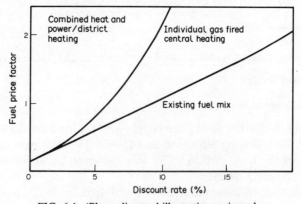

FIG. 6.1. 'Phase diagram' illustrating region where CHP/District heating is economic (after Hewitt and Owen[4]). Housing density = 20 dwellings/acre. Fuel price in year 2000 = (present fuel price) × (fuel price factor).

be remembered that parallel installation of similar engines often enables reduction in demand to be met simply by closing down one or more engines (see ref. 5).

Developments of CHP/IND in recent years appear to have arisen more from both the customer (industry or local authority) *and* the contractor (usually the engine or turbine manufacturer) rather than from a national utility. The Hereford scheme (section 5.6) is an exception, arising from an initiative by the local (area) board of a national generating utility.

The critical parameters in CHP/IND application are not only thermo-dynamic. Housing density is obviously not relevant, but discount rate, plant life and escalation in fuel prices (and their differentials) are import-ant, as for CHP/DH. However, sale (or buy-back) values and stand-by prices of electricity are now of vital importance (see the work of Kehlhofer (Fig. 4.2 for example)). Also critical are any modifications made to "normal" fuel prices specially for CHP plant (for example gas tariffs for CHP/IND plants are often fixed relative to those of competitive fuels). High utilisation is of major assistance to economic performance since purchase of "extra" heat or electricity is minimised and average prices come down.

Because of the great variety in applications (and correspondingly in heat and power demands and their ratio) it is not possible to say that any one system is the best for industrial CHP. However it does appear that both small back pressure steam turbines and gas turbines are gaining wide acceptance in the 3–50 MW (electrical) range, and diesel and gas engines at lower levels of power demand. The flexibility of the gas turbine with a heat recovery steam generator, fired or unfired, for process heat (as at Beilen, section 5.4) and even for bottoming steam turbines with extraction for district heating (as at Saarbrücken, section 5.5) is noteworthy.

It is perhaps surprising that heat pumps have not been more widely used, in conjuction with back pressure steam turbines (see section 3.4.4), particularly for CHP schemes with high values of λ_D, and for heat-only applications ($\lambda_D \rightarrow \infty$). Heat pumps have not been developed on a big enough scale for large city district heating, although they are available in smaller sizes. However, while reversible thermodynamic analysis suggests heat pumps driven by conventional power plants should be attractive for quite low heat to power ratios (λ_D), the introduction of realistic (irreversible) performance parameters shifts the minimum possible value of λ_D to much higher levels and suggests application to "heat-only" schemes rather than CHP.

6.4 The Future of CHP

Further implementation of all CHP schemes will depend largely on economic factors, particularly discount rates, and any escalation of fuel prices (with increased emphasis on energy savings). While the future of CHP/DH is likely to wait on the willingness of national governments and/or local authorities to provide the substantial initial capital required, the popularity of industrial CHP/IND schemes will grow particularly if fuel prices increase. The thermodynamics of CHP is straightforward, the technology is available and its widespread adoption rests on economic factors.

References

1. Combined Heat and Electrical Power Generation in the United Kingdom (the Marshall Report). Department of Energy, Energy Paper 35, HMSO, 1979.
2. District Heating Combined with Electricity Generation in the United Kingdom. Department of Energy, Energy Paper 20, HMSO, 1977.
3. Macadam, J. A., Jebson, D. A., Owen, R. G. and Brogan, R. J. District Heating Combined with Electricity Generation: a Study of Some of the Factors Which Influence Cost Effectiveness. Department of Energy Paper (April 1981).
4. Hewitt, G. F. and Owen, R. G. The Potential of Combined Heat and Power Generation for Space and Water Heating in the UK. CIBS/University of Bath Symposium, Energy, Services and Buildings, 1980.
5. Hein, K. Local Power and Heat Generation in Heidenheim and West Germany. Local Heat and Power Generation: A New Opportunity for British Industry (pp. 89–97). Intersciences Enterprises Limited, Jersey, 1983.

Some Results of Thermodynamic Availability Theory

A.1 Introduction

The theorems of thermodynamic availability, and their applications, are presented comprehensively by Haywood[1] and the "exergy" methods of analysing thermal power plant has been described by Kotas[2]. Here we summarise briefly the results of direct relevance to the steady-flow analyses of plant performance which are presented in the main text.

A.2 Work Output in a Reversible Flow Process

Figure A.1 (after Haywood[1]) shows a fully reversible steady-flow process, between specified stable states (1 and 2) of a fluid, in the presence of an environment at temperature T_0, pressure p_0. The fluid flows through the rigid control surface CV, and rejects heat reversibly at temperature T_U. Work $[(W_{CV})_{REV}]_1^2$ is delivered from CV, and reversible auxiliary cyclic heat engines, operating between temperature T_U and temperature T_0 deliver "external" work $[(W_e)_{REV}]_1^2$. The total or gross work delivered from the extended (dotted) control surface is

$$[W_{REV}]_1^2 = [(W_{CV})_{REV}]_1^2 + [(W_e)_{REV}]_1^2. \qquad (A.1)$$

Availability theory shows that this is the maximum "shaft" work that can be delivered in steady flow between stable states 1 and 2 in the presence of the conceptual environment at temperature T_0, and is given by

$$[W_{REV}]_1^2 = B_1 - B_2, \qquad (A.2)$$

where $B = H - T_0 S$ is the steady flow availability function. The reversible work output between a state 1 and a so-called dead state (at p_0, T_0) is therefore

$$E = B_1 - B_0, \qquad (A.3)$$

where E is termed the exergy.

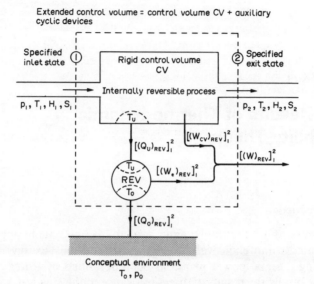

FIG. A.1. Fully reversible steady flow process between specified stable states of a fluid, in the presence of a specified environment.

Thus the work output between states 1 and 2 may also be written as

$$[W_{\text{REV}}]_1^2 = (B_1 - B_0) - (B_2 - B_0) = E_1 - E_2. \tag{A.4}$$

Horlock and Haywood[3] discuss the case arising in the study of CHP plant where heat $\int_1^2 dQ_U$ is rejected at a variable temperature T_U from the control volume CV and is used for useful heating purposes, rather than as a source of heat to the auxiliary cyclic heat engines. The maximum (reversible) work obtained from the "inner" control volume CV is then equal to

$$[(W_{\text{CV}})_{\text{REV}}]_1^2 = [W_{\text{REV}}]_1^2 - [(W_e)_{\text{REV}}]_1^2. \tag{A.5}$$

But

$$[(W_e)_{\text{REV}}]_1^2 = \int_1^2 \left(\frac{T_U - T_0}{T_U}\right) dQ_U, \tag{A.6}$$

so that
$$[(W_{CV})_{REV}]_1^2 = [W_{REV}]_1^2 = \int_1^2 \left(\frac{T_U - T_0}{T_U}\right) dQ_U$$

$$= (B_1 - B_2) - \int_1^2 \left(\frac{T_U - T_0}{T_U}\right) dQ_U \qquad \text{(A.7a)}$$

$$= (E_1 - E_2) - \int_1^2 \int \frac{T_U - T_0}{T_U} dQ_U. \qquad \text{(A.7b)}$$

The reduction in work due to heat transfer is the integral $\int_1^2 \left(\frac{T_U - T_0}{T_U}\right) dQ_U$ so that a Z-factor for this reversible process, due to heat transfer dQ_U, is

$$Z = \frac{\text{Lost Work}}{\text{Useful Heat Transferred}} = \frac{\int_1^2 \left(\frac{T_U - T_0}{T_U}\right) dQ_U}{\int_1^2 dQ_U}. \qquad \text{(A.8)}$$

(See section 2.6.2.)

If T_U is constant, then

$$Z = \frac{T_U - T_0}{T_U} = \eta_{CAR}, \qquad \text{(A.9)}$$

where η_{CAR} is the thermal efficiency of a Carnot engine operating between T_U and T_0.

If T_U is variable then

$$Z = \frac{\int_1^2 \eta_{CAR} \, dQ_U}{\int_1^2 dQ_U}, \qquad \text{(A.10)}$$

and is the "weighted" Carnot efficiency over the range of temperature T_U (see Horlock[4]).

A.3 Work Output in a Real Irreversible Flow Process

In a real (irreversible) flow process through a control volume (CV) between fluid states 1 and 2 (Fig. A.2), heat is rejected at temperature T_U and the work output is $[W_{CV}]_1^2$.

From the steady-flow energy equation

$$[W_{CV}]_1^2 = H_1 - H_2 - Q_U. \qquad \text{(A.11)}$$

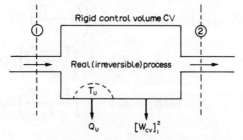

FIG. A.2. Real (irreversible) process
with heat rejection (after
Horlock and Haywood[3]).

From equations (A7a) and (A11),

$$[(W_{CV})_{REV}]_1^2 - [W_{CV}]_1^2$$

$$= (B_1 - B_2) - \int_1^2 \left(\frac{T_U - T_0}{T_U}\right) dQ_U - (H_1 - H_2) + Q_U$$

$$= T_0(S_2 - S_1) + \int_1^2 \left(\frac{T_0}{T_U}\right) dQ_U. \tag{A.12}$$

Generally, the heat rejection (Q_U) does not necessarily occur throughout the process between states 1 and 2.

The entropy leaving the control volume due to the heat rejection is $(\Delta S_U)_{1-2} = \int_1^2 \frac{dQ_U}{T_U}$ where T_U is the temperature at the point where dQ_U occurs. The entropy creation (ΔS_{CV}) due to irreversibility within the control surface CV is by definition (see Haywood[1] p. 296) that part of the entropy change which cannot be accounted for by the entropy transfer associated with the heat transfer. Hence

$$S_2 - S_1 = \Delta S_{CV} - (\Delta S_U)_{1-2},$$

and

$$T_0 \Delta S_{CV} = T_0(S_2 - S_1) + T_0 \int_1^2 \frac{dQ_U}{T_U}. \tag{A.13}$$

From equations (A.12) and (A.13) it follows that

$$[(W_{CV})_{REV} - W_{CV}]_1^2 = T_0 \Delta S_{CV} = I, \tag{A.14}$$

and this is the lost work due to irreversibility within the control volume (CV). Hence the actual work output between states 1 and 2 is

$$[W_{CV}]_1^2 = [(W_{CV})_{REV}]_1^2 - T_0 \Delta S_{CV}$$

$$= (B_1 - B_2) - \int_1^2 \left(\frac{T_U - T_0}{T_U} \right) dQ_U - I \qquad (A.15a)$$

$$= (E_1 - E_2) - \int_1^2 \left(\frac{T_U - T_0}{T_U} \right) dQ_U - I. \qquad (A.15b)$$

(Maximum reversible   (Reduction in maximum   (Lost work due
work without useful   reversible work due to   to irreversibility)
heat rejection)   useful heat rejection)

The Z-factor is now

$$Z = \left[\int_1^2 \left(\frac{T_U - T_0}{T_U} \right) dQ_U + T_0 \Delta S_{CV} \right] \bigg/ \int_1^2 dQ_U. \qquad (A.16)$$

Thus in the real process the Z factor is greater than for the reversible process, when Z is equal to the "weighted" Carnot efficiency (the minimum Z that it is possible to achieve).

A.4   Work Output in a Chemical Reaction at (p_0, T_0)

The (maximum) reversible work in steady flow between reactants at an entry state $R_0(p_0, T_0)$ and products at a leaving state $P_0(p_0, T_0)$ is

$$[(W_{CV})_{REV}]_1^2 = (B_R)_0 - (B_P)_0 \qquad (A.17)$$

(it is supposed here that the various reactants are separated at p_0, T_0; similarly for the various products). The maximum work may then be written

$$(B_R)_0 - (B_P)_0 = (G_R)_0 - (G_P)_0 = -\Delta G_0, \qquad (A.18)$$

where G is the Gibbs function, $G = H - TS$. This is the maximum work obtainable from such a combustion process, and is usually used in defining the rational efficiency of open-circuit plant. It should be noted that if reactants and/or products are mixed at a total pressure of p_0, then work of delivery or extraction has to be allowed for, and the expression for maximum work has to be modified (see refs 1 and 2).

A.5 Comparison Between Maximum Work Output in Chemical Reaction and Actual Work Output

The statements on work output made for a real process (section A.3) and the ideal chemical reaction or combustion process at p_0, T_0 (section A.4) may be compared:

$$B_1 = B_2 + [W_{CV}]_1^2 + \int_1^2 \left(\frac{T_U - T_0}{T_U}\right) dQ_U + T_0 \Delta S_{CV}; \qquad (A.19)$$

$$(G_R)_0 = (G_P)_0 + [(W_{CV})_{REV}]_{R_0}^{P_0}. \qquad (A.20)$$

A.5.1 Open Circuit Plant

If the first equation (A.19) is thought of as applied to an open-circuit CHP plant receiving reactants at state R_1 and discharging products at state P_2 (Fig. A.3a), then additional reversible work transfer processes may be introduced upstream and downstream of the plant to change the reactant flux from $(G_R)_0$ to $(B_R)_1$, and the product flux from $(B_P)_2$ to $(G_P)_0$. Equation (A.19) may be written (subtracting $(G_R)_0$ from each side),

$$(B_R)_1 - (G_R)_0 = ((B_P)_2 - (G_P)_0) + [W_{CV}]_1^2$$

$$+ \int_1^2 \left(\frac{T_U - T_0}{T_U}\right) dQ_U + T_0 \Delta S_{CV} + ((G_P)_0 - (G_R)_0); \qquad (A.21)$$

or rearranging, and introducing (A.20),

$$(G_R)_0 - (G_P)_0 = [(W_{CV})_{REV}]_{R_0}^{P_0}$$

$$= \underset{1}{[(W_{CV})]_1^2} + \underset{2}{\int_1^2 \left(\frac{T_U - T_0}{T_U}\right) dQ_U} + \underset{3}{T_0 \Delta S_{CV}}$$

$$+ \underset{4}{((B_P)_2 - (G_P)_0)} - \underset{5}{((B_R)_1 - (G_R)_0)}. \qquad (A.22)$$

This equation shows clearly how the maximum work output $((G_R)_0 - (G_P)_0)$ splits down into the terms on the right-hand side:

(1) the actual work output;
(2) the work potential of useful heat rejected;
(3) lost work due to internal irreversibility;
(4) the work potential of the discharge exhaust gases.

The last term (5) will usually be zero since reactants will enter at ambient temperature and $(B_R)_1 = (G_R)_0$.

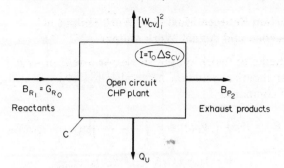

(a) Availability flow, heat and work output

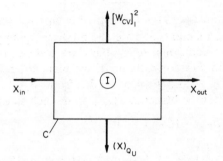

(b) Exergy balance

FIG. A.3. Open circuit CHP plant.

Equation (A.22) may also be interpreted in terms of exergy flows, where $X = (B - B_0) = (B - G_0)$. Thus if $(B_R)_1 = (G_R)_0$ $((X_R)_1 = 0)$,

$$X_{IN} = \text{(Work Output)} + (X)_{Q_U} + I + X_{OUT}, \qquad \text{(A.23)}$$

where $X_{IN} = (G_R)_0 - (G_P)_0 = -\Delta G_0$;

$$(X)_{Q_U} = \int \left(\frac{T_U - T_0}{T_U} \right) dQ_U;$$

$$X_{OUT} = (B_P)_2 - (G_P)_0$$

(see Fig. A.3(b)). This is essentially the approach adopted by Kotas[2] and others in using exergy analysis.

A.5.2   Closed Circuit Plant

For a closed circuit CHP plant (control surface Y, Fig. A.4(a)) receiving heat from an external source (control surface Z, in which combustion

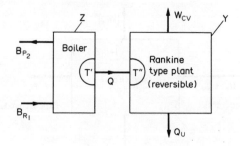

(a) Availability flow, heat and work output

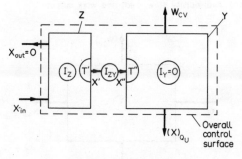

(b) Exergy balance

FIG. A.4. Closed circuit CHP plant.

takes place), equations (A.19) and (A.20) can still apply, but need further interpretation.

Consider for example a back-pressure Rankine plant within control surface Y, from which all heat rejection (Q_U) at temperature T_U is useful, work W_{CV} is delivered and heat Q is received at variable temperature T''. Since the processes within control surface Y are reversible $I_Y = 0$, and the exergy flow for Y (see Fig. A.4(b)) may be written

$$X'' = W_{CV} + (X)_{Q_U}, \qquad (A.24a)$$

i.e.

$$\int \left(\frac{T'' - T_0}{T''}\right) dQ = W_{CV} + \int \left(\frac{T_U - T_0}{T_U}\right) dQ_U \qquad (A.24b)$$

(this is easily visualised from a T,s diagram for the Rankine back-pressure cycle.)

However, the processes within the boiler (control surface Z) are not reversible; nor is the transfer of heat between Z and Y, because of the temperature drop $(T' - T'')$.

For control surface Z, the exergy balance may be written

$$X_{IN} = X_{OUT} + X' + I_Z, \tag{A.25}$$

where I_Z is the irreversibility *within* the boiler (due primarily to the combustion process). Equations (A.24) and (A.25) may be combined to give, for the overall (dotted) control surface of Fig. A.4(b)

$$X_{IN} = X_{OUT} + (X' - X'') + X'' + I_Z$$

$$= X_{OUT} + (X' - X'') + W_{CV} + \int \left(\frac{T_U - T_0}{T_U} \right) dQ_U + I_Z \tag{A.26}$$

Noting that

$$X' = \int \left(\frac{T' - T_0}{T'} \right) dQ, \quad X'' = \int \frac{T'' - T_0}{T''} dQ,$$

it follows that

$$(X' - X'') = T_0 \int \left(\frac{T' - T''}{T'T''} \right) dQ = I_{ZY}, \tag{A.27}$$

the irreversibility involved in transfer of heat Q across the (variable) finite temperature differences $(T' - T'')$ which inevitably occur between the boiler gases and the steam flow.

Further, assuming that reactants enter at ambient conditions (p_0, T_0) it again follows that $(B_R)_1 = (G_R)_0$ and

$$X_{IN} = (G_R)_0 - (G_P)_0 = -\Delta G_0,$$

$$X_{OUT} = (B_P)_2 - (G_P)_0.$$

Thus the final statement of exergy flux for the whole plant becomes

$$-\Delta G_0 = ((B_P)_2 - (G_P)_0) + I_{ZY} + W_{CV}$$

$$+ \int \left(\frac{T_U - T_0}{T_U} \right) dQ_U + I_Z. \tag{A.28}$$

References

1. Haywood, R. W. *Equilibrium Thermodynamics for Engineers and Scientists*. John Wiley, London, 1980.
2. Kotas, T. J. *The Exergy Method of Thermal Plant Analysis*. Butterworths, London, 1985.
3. Horlock, J. H. and Haywood, R. W. Thermodynamic Availability and its Application to Combined Heat and Power Plant. *Proc. Instn. Mech. Engrs.* **199**, C1, 11–17, 1985.
4. Horlock, J. H. *Combined Heat and Power Supply Using Carnot Engines*. Liber Amicorum Andre L. Jaumotte. Note Technique 50 de l'Institute de Mecanique Appliquee, 1983.

Comparative Profitability and Pricing Based on the Time Value of Money

B.1 Introduction

Most assessments of profitability of CHP projects take account of the time value of money. A sum M invested at interest (i) at time zero acquires a value $M(1 + i)$ after 1 year, $M(1 + i)^2$ after 2 years, $M(1 + i)^N$ after N years (Fig. B.1(a)). An alternative interpretation is that a sum M, spent or saved in year N, may be discounted back to year zero, where it has a value of $M/(1 + i)^N$ (Fig. B.1(b)); in this context i becomes a discount rate.

For a CHP project, a capital sum ΔC, invested at time zero, may result in an annual money saving of ΔM in each of years, $1, 2, 3 \ldots k \ldots N$, and invested at the end of each year. Those savings have to be discounted back to time zero for comparison with the sum invested, as Fig. B.2 shows. (By using ΔC and ΔM, we imply that extra capital (ΔC) has been used to convert or change an existing scheme, and that this investment produces savings ΔM in comparison with that original scheme.)

An excellent simple account of comparative profitability based on discounted cash flow (DCF) analysis is given by Holland *et al.*[1] They list many ways of expressing profitability including:

  (i) net present value based on a particular discount rate, NPV;
 (ii) discounted cash flow rate of return, DCFRR, or rate of return on investment, ROI;
(iii) discounted break-even point, DBEP;
 (iv) discounted pay-back period, DPBP.

These are described in section B.2.

In this book, the particular DCF analysis most extensively used is the net present value (NPV)—usually, but not always, of the *changes* from existing plant to a new CHP plant (i.e. capital ΔC invested produces savings ΔM p.a.). The NPV may be determined from computer calculations but simplified analytical results may sometimes be used to give

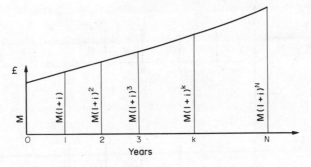

(a) £M invested at time zero

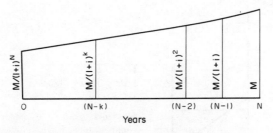

(b) £M discounted back from end of year N

FIG. B.1. Money values.

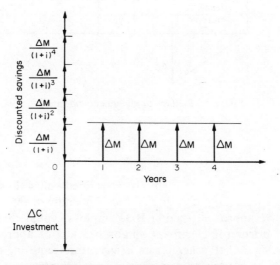

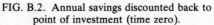

FIG. B.2. Annual savings discounted back to
point of investment (time zero).

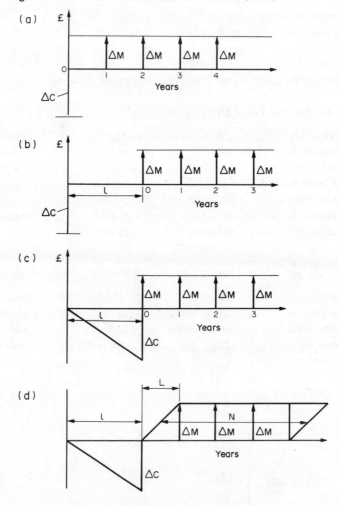

FIG. B.3. Patterns of investment and savings.

approximate answers, and these are also described below, in section B.2. (These analyses were originally given in an Appendix of ref. 2.)

An alternative approach (section B.3) employs capital charge rates in the setting of prices (of electricity and/or heat). This approach usually involves assessing a CHP scheme as a *new* venture requiring total capital C, and pricing the electricity, and/or heat to take account of fuel, maintenance, capital and other costs, including taxes. The resulting prices are then compared with those of an existing plant. The time value of money is again implicit in this approach.

B.2   Discounted Cash Flow Techniques—Simple Analyses

B.2.1   Net Present Value (NPV)

We take the NPV of a CHP scheme as the amount of money that is saved year by year (savings principally due to lower fuel costs, less capital spent) invested up to the end of the plant life of N years, but then discounted back to the start of the whole operation. NPV thus depends on the distribution of capital input and money saving with time. Various distributions are shown in Fig. B.3; savings (full or part) begin at year zero, but capital may be invested before that time.

B.2.1.1   Single Capital Input Leading to Immediate Savings

The simplest case is that shown in Fig. B.3(a) where capital £ΔC is spent at time zero and savings (based mainly on reduced fuel costs) are immediately achieved. They are constant (£ΔM each year), and invested at the *end* of each year. Then the net present value at time zero of the savings over a period of N years is

$$(NPV)_{(i)} = \Delta M \left[\frac{1}{(1 + i)} + \frac{1}{(1 + i)^2} \cdots + \frac{1}{(1 + i)^N} \right] - \Delta C$$

$$= \Delta M \left[\sum_{k=1}^{N} \frac{1}{(1 + i)^k} \right] - \Delta C$$

$$= \frac{\Delta M}{_N f_{AP}} - \Delta C, \tag{B.1}$$

where

$$_N f_{AP} = \frac{(1 + i)^N}{[1 + (1 + i) + (1 + i)^2 + \ldots (1 + i)^{N-1}]} = \frac{i(1 + i)^N}{[(1 + i)^N - 1]} \tag{B.2}$$

is the annuity present worth factor, which may be obtained from tables if the discount or interest rate (i) and the period N years are both known (see Holland *et al.* (Ref. 1)).

B.2.1.2   Single Capital Input Leading to Deferred Savings

There may be a delay of l years before the annual savings of £ΔM begin, i.e. the capital investment ΔC is made l years ahead of time zero (Fig.

B.3(b)). The value of the savings discounted back to time zero is $\Delta M / _\text{N}f_\text{AP}$ when the value of the capital is $\Delta C(1 + i)^l$. The net present value at time zero is therefore

$$(\text{NPV})_{\text{(ii)}} = \frac{\Delta M}{_\text{N}f_\text{AP}} - \Delta C(1 + i)^l. \tag{B.3}$$

B.2.1.3   Linear Spend of Capital

Figure B.3(c) shows the case where the capital is spent linearly over L years ahead of the year zero (annual spend, $(\Delta C/L)$). Money savings ΔM still begin at year zero. Then at time zero

$$(\text{NPV})_{\text{(iii)}} = \frac{\Delta M}{_\text{N}f_\text{AP}} - \left(\frac{\Delta C}{L}\right)[1 + (1 + i) + \ldots (1 + i)^{L-1}]$$

$$= \frac{\Delta M}{_\text{N}f_\text{AP}} - \frac{\Delta C}{L(_\text{L}f_\text{AF})}, \tag{B.4}$$

where

$$_\text{L}f_\text{AF} = \frac{1}{[1 + (1 + i) + (1 + i)^2 + \ldots (1 + i)^{L-1}]} = \frac{i}{[(1 + i)^L - 1]} \tag{B.5}$$

is the annuity *future* worth factor over L years.

B.2.1.4   Linear Spend of Capital and Linear Growth of Savings

A more general case is shown in Fig. B.3(d) where annual money savings also build up to ΔM over L years, starting from time zero, as houses are progressively converted in a district heating scheme. The capital input is also linear over L years but starts l years ahead of time zero.

Consider a year within the building conversion period of L years. The money savings on houses converted in this year are $\left(\dfrac{\Delta M}{L}\right)$ and this carries forward in the following years over a period of N years, for that is the time the conversion lasts. The NPV of these savings at the end of the year under consideration is $\Delta M/[L(_\text{N}f_\text{AP})]$ and this may be discounted back to time zero.

The sum of the discounted savings in all L years at time zero is then

$$\frac{\Delta M}{L(_\mathrm{N}f_\mathrm{AP})} \left[\frac{1}{(1+i)} + \frac{1}{(1+i)^2} + \cdots \frac{1}{(1+i)^L} \right]$$

$$= \frac{\Delta M}{L(_\mathrm{N}f_\mathrm{AP})} \left[\frac{1 + (1+i) + (1+i)^2 \ldots + (1+i)^{L-1}}{(1+i)^L} \right]$$

$$= \frac{\Delta M}{L(_\mathrm{N}f_\mathrm{AP})(_\mathrm{L}f_\mathrm{AP})}.$$

The capital spend over L years ($\Delta C/L$ per annum) has a lead time of l years. The NPV at the start of spending is $\Delta C/L(_\mathrm{L}f_\mathrm{AP})$, but with interest i over l years the NPV at time zero is

$$\frac{\Delta C(1+i)^l}{L(_\mathrm{L}f_\mathrm{AP})}.$$

Thus the total NPV for the operation at time zero is

$$(\mathrm{NPV})_{(iv)} = \frac{1}{L(_\mathrm{L}f_\mathrm{AP})} \left[\frac{\Delta M}{(_\mathrm{N}f_\mathrm{AP})} - \Delta C(1+i)^l \right]. \tag{B.6}$$

If there is no lead, so that capital spend and savings are in phase, then

$$(\mathrm{NPV})_{(iv)} = \frac{1}{L(_\mathrm{L}f_\mathrm{AP})} \left[\frac{\Delta M}{_\mathrm{N}f_\mathrm{AP}} - \Delta C \right] = \frac{(\mathrm{NPV})_{(i)}}{L(_\mathrm{L}f_\mathrm{AP})}. \tag{B.7}$$

In some examples of linear growth the quantity in square brackets is referred to as the NPV "per discounted house". This quantity is the NPV given by assuming all the capital (ΔC) is invested at time zero, and the savings (ΔM) occur immediately (the first example given in section B.2.1).

B.2.2   Net Present Value–More General Determination

The first three analyses of (NPV) given in section B.2 are based on constant savings (ΔM) in each year and, in the last of these three examples, on constant capital investment ($\Delta C/L$) in each year. The cash flow in one year (in this third example) is $A_\mathrm{CF} = A_\mathrm{I} - A_\mathrm{C}$ where $A_\mathrm{I} = \Delta M$, $A_\mathrm{C} = \Delta C/L$. More generally there is a cash flow (A_CF) in any year,

$$A_\mathrm{CF} = A_\mathrm{I} - A_\mathrm{t} - A_\mathrm{C}, \tag{B.8}$$

where A_I is the net annual cash income (or savings);

  A_t is the income tax paid;

and A_C is the capital expenditure.

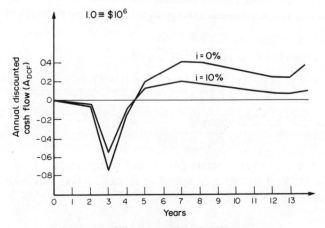

FIG. B.4. Discounted cash flow against time
(after Holland *et al.*[1]).

If this cash flow occurs in the kth year, then it may be invested up to the end of N years and discounted back to year zero,

$$(A_{DCF})_k = \frac{(A_{CF})_k (1 + i)^{N-k}}{(1 + i)^N} = \frac{A_{CF}}{(1 + i)^k}. \qquad (B.9)$$

As in the analyses of the last section, the sum of the discounted cash flow $\left(\sum_{k=1}^{N} (A_{DCF})_k \right)$ gives the net present value

$$NPV = \sum_{k=1}^{N} (A_{DCF})_k = \sum_{k=1}^{N} \frac{(A_{CF})_k}{(1 + i)^k}. \qquad (B.10)$$

This more general statement allows the cash income and capital expenditure to vary year by year. Figure B.4 (taken from Holland *et al.*[1]) shows annual cash flows for a particular project ($A_{CF} = A_{DCF}$ ($i = 0$)) together with discounted cash flows year by year (A_{DCF} ($i = 0.1$)). The second set of data enables (NPV) to be determined by summation.

B.2.2.1 Discounted Cash Flow Rate of Return (DCFRR, ROI)

The interest (or discount) rate (i) can be chosen to make (NPV) zero after a chosen number of years, N. This value of i, given by

$$NPV = \sum_{k=1}^{N} (A_{DCF})_k = \sum_{k=1}^{N} \frac{A_{CF}}{(1 + i)^k} = 0, \qquad (B.11)$$

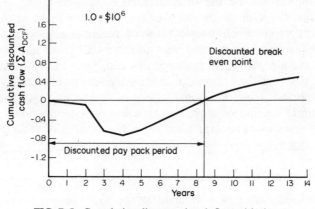

FIG. B.5. Cumulative discounted cash flow with time
(after Holland *et al.*[1]).

is known as the discounted cash flow rate of return (DCFRR). It is sometimes also called the profitability index, the investor's rate of return (ROI),
or the interest rate of return, and is widely used.

B.2.2.2   Discounted Break-even Point (DBEP) and
Discounted Pay-back Period (DPBP)

The discounted break-even point is based on a *selected* rate of interest
(i) and occurs where a cumulative discounted cash flow curve (see Fig. B.5,
which is based on Fig. B.4) crosses the zero line. At this point the project
has repaid the capital and produced the same rate of return on equivalent
capital invested at the selected rate (i). It occurs after a number of years
equal to the discounted pay-back period (DPBP).

B.2.2.3   Comparison with Traditional Simple Methods of
Assessment

More traditional assessments of profitability are often based on simple
rate of return, for example the percentage rate of return

$$\left(\frac{\text{Annual return}}{\text{Invested capital}} \times 100 \right)$$

or a simple pay-back period i.e. the number of years required to pay back
the fixed capital investment from accumulated cash flow. However, there
are many variations of these simple concepts. For example the rate of
return can be based on the annual cash income, or on profit after tax.

Invested capital can be the original total capital cost, the depreciated capital value, the average capital over the life of the plant or the current replacement value. Such simple rates of return and pay-back periods clearly require careful definition, and do not usually take into account the period before a project generates positive cash flow. However, if the time scales are short, as they often are in industry which requires a very short pay-back period, the "traditional" rates of return and pay-back period do not require the use of more complex discounted cash flows. The parameters of assessment become identical as the number of years decreases.

B.2.3 Levelised Cash Flows

The levelised cash flow is defined as the hypothetical (constant) cash flow which gives the same net value as the real (variable) cash flow for the same interest (i) and the same plant life (N years).

Variable cash flows invested at the end of each of several years accumulate interest up to the end of the life of the plant (N years), but, when discounted back to time zero, their sum is

$$\frac{1}{(1 + i)^N} \sum_{k=1}^{N} (A_{CF})_k (1 + i)^{(N-k)},$$

where k refers to the kth year in a life of N years. This is the net present value of all the (variable) cash flows $(A_{CF})_k$, given in equation (B.10),

$$NPV = \sum_{k=1}^{N} \frac{(A_{CF})_k}{(1 + i)^k}. \tag{B.10}$$

If the cash flow is constant ($A_{CF} = \bar{A}$) then

$$(\overline{NPV}) = \bar{A} \sum_{k=1}^{N} \frac{1}{(1 + i)^k} = \frac{\bar{A}}{(_N f_{AP})}, \tag{B.12}$$

where $(_N f_{AP})$ is again the annuity present worth factor,

$$(_N f_{AP}) = \frac{i(1 + i)^N}{(1 + i)^N - 1}. \tag{B.2}$$

If the two net present values (NPV) and $(\overline{NPV})$ are to be identical then

$$NPV = \sum_{k=1}^{N} \frac{(A_{CF})_k}{(1 + i)^k} = (\overline{NPV}) = \frac{\bar{A}}{_N f_{AP}}. \tag{B.13}$$

Thus the levelised cash flow is

$$\bar{A} = (_N f_{AP}) \sum_{k=1}^{N} \frac{A_{CF}}{(1 + i)^k} = (_N f_{AP})(NPV). \tag{B.14}$$

B.3   Pricing of Heat or Electricity

In this section we consider a different approach—that of pricing the product to take account of annual fuel or maintenance costs *and* capital costs. The concept of capitalised cost of replacement is first examined.

B.3.1   Pricing to Take Account of Capitalised Cost

Suppose the capital cost of a CHP plant in year zero is fixed at C_0, and the plant has a life of N years. A capital sum $C > C_0$ must be provided at time zero, so that after N years a sum of $(C - C_0)(1 + i)^N$ is available to replace the plant. If the salvage value of the plant after N years is (SV) then the total replacement cost (plus provision for future renewal) is $C - $ (SV) (assuming zero inflation). These two quantities must be equal if the CHP plant is to be renewed indefinitely, i.e.

$$(C - C_0)(1 + i)^N = C - (SV),$$

and

$$C = \left[C_0 - \frac{(SV)}{(1 + i)^N} \right] \frac{(1 + i)^N}{[(1 + i)^N - 1]}$$

$$= \left[C_0 - \frac{(SV)}{(1 + i)^N} \right] f_N, \tag{B.15}$$

where

$$f_N = \frac{(1 + i)^N}{(1 + i)^N - 1} = \frac{(_N f_{AP})}{i} \tag{B.16}$$

is called the capitalised cost factor.

If the salvage value (SV) $= 0$ then the capitalised cost C is

$$C = C_0(f_N) = \frac{C_0(_N f_{AP})}{(i)}. \tag{B.17}$$

The interest obtained on capital C if it were used elsewhere would be

$$Ci = C_0(f_N)i = C_0(_N f_{AP}). \tag{B.18}$$

Thus an investor in the project will wish to have his capital C preserved after N years, and to receive interest (Ci) annually. These requirements are met if the project pays an annual capital charge of $C_0(_N f_{AP})$.

B.3.2   Capital Charge Rates

Thus in methods of assessment involving pricing of electricity and/or heat, account must be taken of these capital costs. The annual capital charge rate must be added on to other costs of production (fuel, operation and maintenance charges).

For a *new scheme* (conventional or CHP), with fuel, maintenance and other costs of £M p.a. requiring capital £C_0 invested at time zero capitalised cost £C) the total costs will be

$$M + Ci = M + C_0\beta,$$

where
$$\beta = (_\text{N}f_\text{AP}) = \frac{i(1 + i)^N}{(1 + i)^N - 1} = if_\text{N} \qquad (B.19)$$

is the familiar annuity factor (sometimes in this approach called the capital recovery factor). The product (electricity and/or heat) must be priced to meet these costs, plus taxes, and provide a profit.

Williams[3] uses this method, but elaborates it to take account of many other factors beyond the simple interest rate. He defines i as the discount rate

$$i = \alpha_e r_e + (1 - t)\alpha_d r_d, \qquad (B.20)$$

where α_e, α_d are the fractions of investment from equity and debt, r_e, r_d are the corresponding annual rates of return, and t is the income tax rate. Allowance may also be made for any investment tax credit rates (t_c) insurance rate (r_i) and capital replacement rate (r_r), assuming straight line depreciation. The capital recovery factor used by Williams is then

$$\beta' = \frac{\beta}{(1 - t)}\left[1 - \frac{t_c}{(1 + i)} - t\right] + r_i + r_r. \qquad (B.21)$$

In using this method of pricing the product, it is usual to regard a CHP scheme as a *new* one, to determine the corresponding prices of heat and/or electricity, and to compare the prices with those required in a conventional plant (see section 4.2).

B.4 Effect of Inflation

Inflation can be allowed for in selecting a discount rate, or rate of return. This would involve setting the rate of return allowing for inflation correspondingly higher than that with zero inflation. For example Williams[3] points out that the rates of return (with zero inflation) used in his studies ($r_e = 0.055$, $r_d = 0.028$) would correspond to $r_e = 0.12$ and $r_d = 0.09$ with 6% inflation.

It is more usual in DCF analyses to assume that overall inflation does not take place and to set the discount rate at the lower levels. However, allowance is then made for *relative* inflation (e.g. a rate of escalation in fuel costs) greater than the overall inflation rate, as in the Marshall report.[2]

References

1. Holland, F. A., Watson, F. A. and Wilkinson, J. K. *Introduction to Process Economics*. John Wiley, London, 1974.
2. Horlock, J. H. Analysis of the Energy and Money Savings Associated with CHP Plant. District Heating Combined with Electricity Generation in the United Kingdom (Appendix 12). Department of Energy, Paper No. 20. HMSO, London, 1977.
3. Williams, R. H. Industrial Cogeneration. *Ann. Review Energy*, **3**, 313–356, 1978.

Index